American Presidents at War

American Presidents at War

How U.S. Commanders-in-Chief Dealt With Wartime Conflicts During Their Years in Office, From the Revolutionary Conflict to the 21st Century War on Terror

By Thomas P. Athridge

Revised Edition, 2018

Cover by John Mulcahy, aois21 media.

Printed by Lulu.com

ISBN: 978-1-941771-25-9

This book is dedicated to the memory of those who lost their lives, and their loved ones, during the September 11th, 2001 terrorist attacks upon the United States. We will always grieve the gross injustice of their loss with you, and stand by you as we try to recover from its devastation.

This book is also dedicated to the U.S. Armed Forces throughout all of American history, including in present military campaigns, some of whom gave their very lives to create the freedom and prosperity that we enjoy today.

Finally, I dedicate this book to all American presidents, past and present, whom this nation entrusted to protect us from the evil entities around the globe that hate our way of life, and the roles that they have played not only with Congress, but also with American history.

In Memoriam

To Thomas P. Athridge, Jr. (1935-2016), who showed me that nothing is impossible, no matter how powerful the adversary, as long as I never give up.

Acknowledgments

I am deeply grateful to the following people, whose influence helped me tremendously in completing this book, which took me fifteen years to accomplish (May 2002 to May 2017): in no particular order, Dr. Allan Hunter, Dr. Robert Keighton, D.L. Garren, Michelle Krowl, Jon Powell, Dr. Louis Fisher, John Sellers, Jerry Gawalt, Everett Larson, Dr. Joann Moran-Cruz, Kimberly Peach, Edward Kader, Ted Westervelt, Kasele Myers, Ralph Cole, Cecelia Desmond, Timothy O. Desmond, Thomas P. Athridge Jr. (Dad), Mary, John, and Mary K. Athridge, John Mobille, Esq., Steven Dahl, Joseph Whalen, Setor Awuyno-Akaba, Richard Ground, Ed Ground, Ed Rosenthal, the staffs of both the Serial Record Division and United States Acquisitions at the Library of Congress, Linda Malone, Bela Saxonov, Sabrina Hsu, James and Jessica Ritter, Jason and Caroline Curry, William and Susan Stone, Mark and Shara Dahl, Dr. Robert Remini, Debra Graham, Cheryl King, Lakeia Roseboro, Lem Gebrekristos, Kimantha Maye Hargesheimer, Liah Love, Lydia Shvdsky, Janice Henson, Leslie Maron, members and officers of Local Union 2477 and 2910 at the Library of Congress, the Asian-American club at the Library of Congress, the Veterans History Project at the Library of Congress, the staff of the Little Falls Library in Bethesda, Maryland, China Pavilion Restaurant, and the Washington Metropolitan Area Transportation system (the subway), where I conducted a vast amount of research for this project by reading on the train.

Why Did I Write This Book?

I chose to research and write this book because of the events of terror that happened in America on September 11th, 2001. I felt so angry, frustrated, hurt, upset, terrorized, sickened, and devastated to witness these evil attacks on innocent civilians on that tragic day that I felt that I personally had to do something in response. I had never before experienced the kind of event of national devastation that America suffered on that day, and it is one that I will never forget. The shock of the devastation of watching the hijacked planes crash into the World Trade Center towers, and then watching their collapse, still haunts me to this very day.

In this project, I was seeking to point out other similar episodes in American history when our nation was dealt similar acts of warfare, in which our federal government enacted our military capabilities to lead our brave soldiers into battle to defeat foreign aggressors. In this research, I was seeking answers to our current state of warfare from leaders of past administrations from both political parties, who were involved in their own war campaigns that circumstances drew them into, and I chose to examine the actions and solutions that they each undertook to reach a successful or unsuccessful conclusion. This idea led me to formulate my analysis to include every war conflict in American history and draw my conclusions based on both the outcomes of military campaigns and individual presidents' performances during their years in office. Each challenge was unique, and difficulties arose in each situation that each executive faced during his individual crisis. In my research for this project, I have learned that our nation has been severely tested by former adversaries, which almost led to the destruction of our free and democratic government. This includes those of our allies abroad, also. In each of these challenges, the United States and our then-allies and present allies have stood up to fight persecution and tyranny occurring at home and around the world. Those past successes gave our society the sense of contentment and security that was shattered on September 11th, 2001.

I want to find meaning in and honor those who not only gave their lives in the cause of fighting for freedom throughout American history, but also all of the innocent lives that were taken on September 11th, 2001. For it was on that fateful day when the United States was once again called into war against an enemy such as we had never faced before, and our leaders would be truly tested to respond overwhelmingly against an enemy that had attacked civilians on our very soil.

Table of Contents

Chapter 1
The Current 21ˢᵗ-Century War and the Media Coverage of It

The world is a very different place in the 21ˢᵗ century. Living off the technological marvels of the 20ᵗʰ century, the United States and our United Nations coalition allies, such as N.A.T.O., can focus on a worldly dilemma in a matter of minutes or hours. Previously, these communications systems took days or weeks during war campaigns. Through all of America's previous military conflicts, dating back to President Washington's victory over the heavily favored British in the Revolution campaign, America has risen to world superpower status. This would include America's survival during a brutal civil war campaign between the northern and southern states, which almost split this nation into two separate entities during the 19ᵗʰ century, stemming from the evil practice of slavery.

In the 20ᵗʰ century, the United States, along with our global allies, defeated Germany in two world wars, along with Japan and Italy in the second. The Soviet Union and the United States, despite being allies in World War Two, would be involved in a Cold War for almost fifty years after the end of the war, ending with the fall of communism in Russia in the early 1990s.Two wars that are related to the Cold War policy are the Korean and Vietnam wars that America engaged in to stop the spread of communism in Asia and to assist our allies in the region in the pursuit of living in free, democratic societies. These campaigns would come at the sacrifice of thousands of brave men and women's lives. During the War of 1812, the British literally burned down the Capitol building and the White House, which threatened the very liberties that the Patriots fought for 3.5 decades prior. Yet, the important Battle of New Orleans, led by General and later President Andrew Jackson, would change the momentum of military victory back to the United States over a more powerful English military, which I will explore in a later chapter.

Since the end of the Vietnam War, our nation has succeeded in another war in the Persian Gulf, survived political scandals, struggles, and tragedies, and prospered and flourished during these changing times.

1

Though we remember the religious ideological struggles that America became engulfed with, including the kidnapping of American hostages in Iran during the late 1970s and the attacks in Lebanon on U.S. Marines in 1983, we here, as citizens of the United States, felt protected from foreign entanglements and those terror organizations that wish to do us harm internationally. The success of the first Gulf War in the early 1990s and the current bloodshed in the Middle East, including the daily problems and violence that still exist between Israel and Palestine, seem to echo the instability of that chaotic region of the world. That violent world of chaos and disorder seemed far away and distant from America before September 11th, 2001, in which the freedom of ideas and occasional disagreements were exchanged daily and frequently among a wide variety of subjects, and taken for granted. Also, the wealth that the United States has obtained, both economically and militarily, is a far cry from the rest of the world's daily struggles that other nations' people have to live in. This is where war, oppression, brutality, and extreme poverty encapsulate people's very existences. While poverty and extreme violence certainly exist here in America, people who live in different, and poorer, regions of the globe, where chances for real opportunity are fewer, can have less chance for freedom and justice as opposed to our people with our lifestyles. Their world subjects people to far more disorder and persecution when compared to ours.

Before September 11th, 2001, watching terrorism on television and reading about it in the newspapers seemed very distant. Although terrorists have been attacking our soldiers and interests in other lands, such as the aforementioned Beirut terror bombings of U.S. Marines in 1983, and the nefarious U.S. embassy attacks in Nairobi and Tanzania in 1998, and the U.S.S. Cole bombing in 2000, we have seemed auspiciously content here in the United States to proceed with our daily rigors and challenges relatively terror free. We, myself included, took for granted that this nation would be safe from the terror that existed overseas and in other countries far from home. That now-obsolete scenario changed drastically over the course of events that took place on one of the most difficult mornings in American history.

As the world now knows, on September 11th, 2001, four hijacked domestic airliners crashed into both World Trade Center towers, the Pentagon, and a Pennsylvania field, which turned out to be a thwarted attempt by the hijackers at a second target in Washington, D.C. These 19 hijackers managed to take control of four jets and brutally kill approximately 3,000 innocent American civilians with box cutters and Al-Qaeda terrorist training. The ruthless efficiency of their destruction and the damage they inflicted shook America to its very core. The brave men and women of the New York Police Department and fire department gave their very lives that day by simply doing their jobs of trying to save as many lives as they could, which came at the price of their own. This is a sacrifice that no one should ever forget, because they define and encompass what a true hero really is.

People from over 80 different countries were killed in the 9/11 attacks, and the economy, especially in New York, was hit hard. According to *Time* magazine's March 11th, 2002 edition, (pg. 28), the C.I.A. and other intelligence agencies were caught by surprise by the terror attack. "It was an abject intelligence failure," according to a White House official, regarding the attack. According to that issue, the Bush Administration was still bracing for another terrorist attack on America. "We are as vulnerable today as we were on September 10th or 12th, we just know more now," according to former Presidential counselor Karen Hughes. (pg 28, *Time*, 3-11-02 issue) The C.I.A. has been criticized in large part due to the agency's failure to gather human intelligence about foreign enemies, and critics also highlighted the lack of cohesion between law enforcement and federal intelligence agencies.

Because the enemy is a new type of foe, not a nation or an established government but a multi-dimensional terror network operating in numerous countries, our fighting forces must adapt in order to defeat it on a global scale. While the Bush Administration was initially praised for forming the best international union possible to wipe out these savage terrorists, we know that the world and our country are just not completely safe from these demented people, whose mission is to cause as much pain and death as they possibly can, at the cost of their own lives. This is attributed to their extreme fundamentalist views. The terrorists' legal entry into the United States shows the porousness of U.S. borders. Even though two hijackers were on the federal terrorist watch list at the time of the tragedy, they were able to purchase airline tickets with credit cards in their own names with relatively little difficulty. This shows the complexity of the federal government and intelligence centers keeping its eye on every single individual all the time, and the need for federal agencies to be in congruence to thwart future attacks on our civilians. (pg 29, *Time*, 3-11-02).

Fifteen of the 19 terrorists who committed the 9/11 attacks were Saudi Arabian citizens, a country with which the United States does not have hostile relations, either currently or during the time of the attack. The F.B.I. and the C.I.A. had one of the terrorists on a watch list for his presence in a videotaped meeting about the attack on the U.S.S. Cole (pg 29, *Time*). His name was Al-Mihdhar. Though the U.S. authorities began a frantic search to place him into custody, he quietly slipped away and was never found. He was aboard Flight 11, which crashed into the Pentagon, on a ticket purchased by him under his own name, because American Airlines stated that "no government authority informed them that he was on a terror watch list". (pg 29, *Time*). These evildoers exposed a weakness in airport and intelligence security, which should have been tight enough to prevent these horrific events from happening on 9/11. However, due to human error and tragic circumstances, it went unnoticed.

We will continue to feel vulnerable and frightened in the post 9/11 world, although airport and mail security have been understandably vastly tightened. There are 13 semiautonomous agencies in the federal intelligence splintered level that share a computer network. However, due to interaction problems on and before 9/11, collaboration was not in their nature at the time (pg 29, *Time*). While agencies and especially people have adapted to being more cooperative since the 9/11 attacks, we still have a long way to go in both the war against terrorism abroad and the peace at home, where people just get used to their normal daily routines, free from terror attacks. Unfortunately, this has understandably been a problem for people who were psychologically and emotionally affected by the terrorist attacks.

As I am writing this chapter, certainly in the New York City area, the mental devastation is rampant, and the gaping holes where the awesome Twin Towers once stood are a painful reminder of the human atrocity that took place there on September 11th, 2001. For those who saw the terror acts up close, the wounds are far too fresh for closure. David Emge, who worked at the U.S. Customs Service at 6 World Trade, said when describing the 9/11 attacks, "It was an unnatural shake. I hadn't heard or felt anything like that since about 30 years earlier in Vietnam. We got out of the library, into the hallway, and the hallway was bustling with people going through the disaster drill. I'm standing there watching this, and I noticed paper coming straight out of the building, and flying out into the plaza. But there were other things coming straight down. They were universally white at one end, and various colors at the other, and I realized that I was watching people who were forced to make a decision a human can't imagine. I had people say to me, 'Well, you were just in the wrong place at the wrong time,' but that's wrong, we were in the right place at the right time. We were at our jobs, doing what we get paid to do. The only people who were at the wrong place at the wrong time were the bastards who flew the planes into the buildings, and we have to remember that."

At the time that I am writing this story, many of the victims have not been found, and the grieving process for the families has not yet come to an end. According to the article "The Long Goodbye," (*New York Magazine*, March 18th, 2002), people like Mary Maciejewski, who lost her husband in the attacks, meet once a week with a group of women who lost their spouses at the World Trade Center. David Fitzpatrick, who is the most senior member of NYPD's Technical Assistance Response Unit, photographed the burning towers from a police helicopter, and he still goes there to take aerial and ground photos (pg 31, *New York Magazine*). According to Fitzpatrick, there is still a strong feeling of comradeship between police, fire, and rescue workers, but as far as the World Trade

Center site itself goes, "it's never lost its feeling of devastation. Even though I go there so often, it just never goes away" (pg 31, *New York Magazine*).

Jessica Trant, who was a 19-year-old freshman at Pace University in Northport, New York, had to endure the loss of her father Daniel, at Cantor Fitzgerald. She was told originally that he would make it out of the building from a phone call her mother received from him only minutes after the building was hit, while Jessica was returning from a canceled business law class. When she returned home, she found her brother curled up into a ball, and crying on the floor. It was then that the truth of what really happened was revealed to her. What her father really said to her mother on the phone was that he was consumed in smoke and fire, and he probably was not going to make it out, but that he was going to attempt an escape. Seemingly knowing his fate, he told his wife that he loved her and the children, and for her to take good care of them (pg 28, *New York Magazine*, March 18th, 2002).

These stories of extreme tragedy are all too real for those family members of each of the victims of the September 11th, 2001 attacks. This is also true for those people who survived to witness the events unfold right in front of their eyes. According to *New York Magazine* article "Living in the Shadow," by Meryl Gordon, a second wave of anxiety and grief swept the New York area during the six-month anniversary of the disaster, amid reports of a second possible attack on the city. In the offices of an organization called Lifenet, the number of the calls they received increased from approximately 3,000 a month before 9/11 to 6,600 by New Year's Eve, many regarding post-traumatic stress disorder from the terror event. This would suggest that the entire collateral damage to the tri-state area has been extraordinarily devastating (pg 23-24, *New York Magazine*, March 18th, 2002). People are apprehensive about taking New York City subways, or driving over the George Washington Bridge. The 9/11 attacks on our nation even traumatized people to the point of affecting their marital relationships (pg 24, *New York Magazine*, March 18th, 2002.). Everyone affected by this tragedy is understandably consumed with anger and grief, and we want to return to the feeling of safety and security that we felt that we all shared before those horrific attacks.

The industries of psychiatry and psychology have both been radically affected in the volume of patients and ways of treating those who have experienced psychological trauma from the terror attacks, including panic episodes and respiratory trouble. Religious leaders, such as Rabbi Peter Rubenstein, have been dealing with questions about God's role and failure to prevent these attacks, as described in the article, "How Do I Find a Spiritual Center or Connection With God During This Time of Loss?" (pg 26, *New York Magazine*, March 18th, 2002). According to this article, the city of New York seemed to ask itself, "What am I doing with my life?

Should I keep doing it here?" It seems as though it will take a long time, if ever, for New Yorkers to feel completely safe again (pg 26, *New York Magazine*, March 18th, 2002). This would also be the case for those affected in the Pentagon and Shanksville, Pennsylvania tragedies. Consequently, due to the many layoffs and damage to the New York economy, this has taken both a financial toll as well as a mental one. This is especially true for people struggling to find work and pay bills during this time.

According to a May 27th, 2002 *Time* magazine article titled, "How the U.S. Missed the Clues," it described how our intelligence agencies failed to connect four substantial clues that might have prevented the 9/11 attacks and the wars in Iraq and Afghanistan. At an August 6th, 2001 briefing to President Bush, it described a possible terrorist attack operation in America. President Bush requested at that time to be updated on the domestic terror capabilities of the organization that was making this threat (pg 28, *Time*, May 27th, 2002). According to the report, it did contain a sentence or two on hijacking airplanes, but then National Security Adviser Condoleezza Rice insisted that there was no information on flying aircrafts into buildings. However, reports contrary to this suggest that in 1995, Philippine authorities uncovered plans for the mass hijacking of multiple airliners and crashing them into the Pacific Ocean. (pg 28, *Time*, May 27th, 2002) Also, Ramzi Yousef, a convicted terrorist, had revealed under investigation to the authorities that there was a plan to crash a plane into the C.I.A. headquarters in Langley, Virginia. This is the same Yousef who, with his terror partner Abdul Hakim Murad, helped mastermind the World Trade Center bombing in 1993.

Another terror plan was foiled in Paris, France, by French authorities to fly an airplane into the Eiffel Tower by the Algerian Armed Islamic Group. A detailed memo issued in July 2001 describes an alert by F.B.I. agent Kenneth Williams, who reported that Arab men training in Arizona flight schools had possible Al-Qaeda terrorist associations that needed to be immediately investigated. Tragically, this important memo was not investigated in time. According to this article, this was the first of four critical intelligence failures that led to the 9/11 tragedy.

The second mistake was that this memo describing the infiltration of American flight schools by suspected terrorists, submitted by agent Williams, was never forwarded to the Chief of International Terrorism (pg 29. *Time*, May 27th, 2002). The third mistake was that the administration and the F.B.I. were concerned with an attack overseas, which was contrary to the Williams memo, which indicated an attack here. To support this claim, the article describes that the "millennium bomber," Ahmed Ressain, had recently planned to detonate a bomb at Los Angeles International Airport in 2000, which, if successful, could have been a catastrophe equaling or easily surpassing the tragedy of 9/11.

Yet, investigators for the United States believed that an attack from terrorists would come overseas, reminiscent of the attacks on the U.S.S. Cole, or the embassy bombings in Nairobi and Tanzania. By July 2001, the warning bell from intelligence sources was deafening about terrorism. There was an apparent plot to kill President George W. Bush when he visited the G-8 summit with world leaders in Genoa, Italy. When the summit ended without incident, the intelligence community moved to possibilities of attacks of U.S. bases in Belgium and Turkey (pg 30, *Time*, May 27th, 2002). Attention was also focused on France when Djanel Beghal, a Franco-Algerian associate of Al-Qaeda, was picked up by authorities while traveling from Afghanistan to Europe, and indicated a plot to attack the American embassy in Paris. This apparently may have been a technique to draw America's attention away from domestic protection at the time of 9/11, when we were expecting an incident overseas.

The fourth mistake revolves around the arrest of Zacarias Moussaoui in August 2001. He was originally arrested for immigration violations when a flight school informed the F.B.I. of Moussaoui's suspicious activities (pg 31, *Time*, May 27th, 2002). When U.S. officials checked with French authorities on Moussaoui's history, they learned of his known associations with extreme Al-Qaeda organizations, from London to Malaysia. Investigators wanted a national security warrant from the F.B.I. headquarters to search Moussaoui's computer files, but they were turned down due to lack of sufficient evidence that he belonged to a terror group (pg 31, *Time*, May 27th, 2002). Agents did, however, check Airman Flight School in Norman, Oklahoma, and informed the C.I.A. about Moussaoui; the C.I.A. had been running checks on him for foreign intelligence services. Unfortunately, neither the C.I.A. nor F.B.I. ever informed the White House counter-terrorism group of their findings.

To avert these possible future intelligence mishaps, the F.B.I and the C.I.A. acknowledged mistakes in sharing information, and subsequently doubled the staff of the counter-terror center (pg 34, *Time*, May 27th, 2002). Also, former F.B.I. Director Robert Mueller announced that a super squad of Washington-based agents whose sole purpose is to investigate terror would be created, along with the hiring of an additional 2000 agents in the 18 months following the May 2002 report. The anthrax attacks have public health officials committed to upgrading U.S. bio-terrorism defense funding from $2.9 billion in 2002 to $4 billion in 2004.

The article that appeared in *Newsweek*'s May 27th, 2002, edition, titled, "What Went Wrong," by Michael Hirsh and Michael Isikoff, essentially echoes the *Time* article and describes the constant threats that are communicated to the United States without substantial credibility, otherwise known as "chatter" (pg 28, *Newsweek*, May 27th, 2002). However, an intelligence team, led by Bill Kurtz, received credible information of suspected hijackers taking airplane-flying lessons at local Arizona flight schools. They also learned that these men were asking questions about airport security. This information came from the aforementioned agent Ken Williams, which he had sent to his superiors. Sadly, this information was ignored, because his bureau was more concerned about racial profiling, domestic crime, drugs, and child pornography (pg 30, *Newsweek*, May 27th, 2002). However, things had changed on August 6th, 2001, when the President was at Camp David and was specifically informed by members of his Cabinet about terror warnings that were received overseas, and of the possibility of airline hijackings occurring in the United States, which Bush became concerned about.

Perhaps the biggest miscommunication was with the Minnesota-based agents, who learned from Moussaoui that there might be an attack on the World Trade Center. However, the Minnesota team did not know about the Phoenix memo. Also, a few weeks after the Phoenix memo warning, the F.B.I. received word that two men on an F.B.I. watch list named Nawaf Alhazmi and Khalid Al-Mihidhar were in the United States. The agency had traced their path to southern California, but failed to locate them. This was despite the fact that Alhazmi was listed in the phone book, and had a bank account in the area. Both of these men were on American Airlines Flight 77, and had they been caught earlier in the summer, which was possible, this attack might have been prevented.

Even Ahmed Rassan, as previously mentioned in the *Time* article, told investigators from prison that despite his attack on LAX airport in Los Angeles being thwarted, Al-Qaeda still planned attacks before Sept. 11th. 2001 on airports because "an airport is sensitive politically and economically" (pg 32, *Newsweek*, May 27th, 2002). According to this article, all these clues should have had American airports on high alert on September 11th, 2001. This was not the case, however, because the two airlines involved in the hijackings were barely aware of the FAA warnings (pg 33, *Newsweek*, May 27th, 2002).

These same comparisons were implemented during the Pearl Harbor tragedy, in which this country and its leaders also asked the familiar questions: What could have prevented this tragedy ? What went wrong? In that attack on December 7th, 1941, 2,403 American servicemen were killed in a surprise attack that involved this country into the conflict that would become World War 2. In that struggle, when our nation was attacked, President Roosevelt called for this nation to fight our enemy aggressively at home and abroad, which would cost many lives in the process, in order to ensure the prosperity and growing strength of our nation and to help stomp out the genocidal atrocities occurring around the world. The result was an Allied victory (United States, Great Britain, and the Soviet Union) over the Axis powers. (Germany, Japan, and Italy). This war brought freedom to many people's lives, for whom such rights would otherwise not be possible. It is also important to note that the United States has since built strong allied relationships with all three countries that were once our adversaries in that campaign and have become global partners in democratic rights and institutions.

Since 9/11, this nation has experienced loss such as we seem to have never experienced before. These feelings of hurt, guilt, pain, and betrayal will be with many of us for a long time. The question that I ask is: Have we, as a nation, ever experienced pain of this magnitude before? Was the very freedom and liberty that we enjoy today ever in peril and in danger of being defeated by similar enemy forces like the ones we face today? What circumstances have led this nation to go to war in the past against other countries and enemies, and is it similar to the war that we are presently in?

Looking back at our past history, as a nation, I believe that we can find examples of heroism and triumph in the face of colossal adversity, and how the triumph was obtained through military and Presidential leadership. I also will examine what laws and improvements our nation made in fighting these wars of American history that enhanced our liberties and lives. I will include the examination of the war the United States is currently in, located in both Afghanistan and Iraq. These past military and political successes made this country stronger, and gave us the sense of security that this nation got used to up until that awful morning of Tuesday, September 11th, 2001.

On the morning of September 11th, 2001, I was at my work station in the Madison building of the Library of Congress, located in the Capitol Hill region of Washington, D.C. It just seemed like countless other mornings at work on a Tuesday. The weather that day was nice, and it certainly seemed like a typical, uneventful morning. I happened to arrive to work at 7:00 a.m., and I was sitting at my desk doing computer work. I remember being annoyed thinking about the Redskins game the previous Sunday, in which they lost to the San Diego Chargers 30-0 to open the season. I saw advertisements for the first National Book Festival, which had occurred three days previously and at which First Lady Laura Bush and other celebrities joined Dr. James Billington and staff for a day to promote reading, which was a big success. Everything seemed so normal that there was no way that I could have predicted the horrific events that were taking place at the time.

I went on break at 9:00 a.m. to get a soda and a snack, and I was back at my desk at 9:30, completely unaware of the human tragedy taking place at the moment. Before my return to my office in a last stop at the bathroom, I overheard two men talking about those planes that crashed into the towers. I was confused by their statement, so I just minded my own business, washed my hands, and returned to my desk. Suddenly, the mood in my office rapidly became scared and panicked, but I still did not understand the magnitude of what was happening, because I was not near a television, and information started only slowly appearing on the internet.

As I was trying to gather information as to what exactly was going on, my boss informed me that I, along with the rest of the staff, was free to go by using what is called annual leave. All government employees are given a certain amount of annual and sick hours per paycheck. Because I knew that I did not have an abundance of either leave category at the time, I was reluctant to use it if I did not have to. I was still unaware of the severity of the situation. From the chaotic scene that my office was becoming, all that I could gather was that a plane (I did not know then that it was two) had crashed into the World Trade Center, and I was also unaware of one of the Towers having collapsed by that point. As I was debating with myself whether to just sign my leave slip and go home, I was then informed of the equally incredible tragedy that a plane had just crashed into the Pentagon. I went to the window on the fifth floor where you can literally see over the Anacostia and Potomac Rivers into Virginia. It was then that I could see with my own eyes that the black smoke billowing into the sky was coming from the Pentagon, and that, in fact, the United States was under attack for real.

This fact then convinced me to use whatever leave I had left, and to get out of the building and go home as soon as possible. As I was filling out my leave slip to go home, a female Library of Congress officer, who was normally calm and cool, burst into the room and ordered an immediate evacuation of the building, and of Capitol Hill entirely. I knew at that moment that we were at war with whoever did this, and that we, as citizens, were the targets. At this point of this confusing morning, I was still unaware of the severity of the attacks in New York, Washington, and Pennsylvania. As I was walking out of the building, I got the story that two hijacked airliners had hit both World Trade Center towers, but I still could not fathom that they would actually collapse as they did, for I did not dream that it was even possible.

The scene of the mass exodus of government employees from the Capitol Hill area, leaving all at once to escape and go home, was surreal and horrific, and a moment that I will never forget. I knew we were under attack, and I was located across the street from the U.S. Capitol building, looking for the safest way home. My three options for getting to my car at my local subway station in Bethesda were as follows: (1) jump in a cab and pay the extra money to ensure that I made it to my car, which was approximately 8 miles away; (2) go to the closest metro station, Capitol South, which connects service between two different lines and would require me to change subway lines at Metro Center, but would eventually bring me to my car, or (3) take a 10- to 15-minute walk to Union Station, enabling me to avoid switching lines and stay on one line in order to get home.

After careful review of these options, I decided that the walk to Union Station seemed the wisest, because if the terrorists attacked the subway, I could narrow the odds of catastrophe by taking only one metro line as opposed to two, which at the time seemed to be a real possibility. The people walking with me were strangely calm, but there was a high sense of panic in the air. The police in front of the U.S. Supreme Court building ordered us to keep moving and not to look back. The people walking with me at the time were saying that possibly up to ten planes were hijacked, on top of the planes we already knew hit the World Trade Center and the Pentagon. I remembered that my sister worked near the World Trade Center at one time, but due to the confusion and panic at the time, I could not remember if she had said on a recent trip home whether she had left that job or not. I thought she had said she had gotten another job in another building, but I was not 100% sure.

We all managed to get to the subway, where we all packed the train destined for our homes. We all sat in silence on the train, as I am sure that we all pondered the multitude of victims who were not able to go home as we were, and the threat against all of us was still very active. As I made it to my subway stop, I rode the escalator up to the street and walked towards my car at a parking lot nearby. It was there that I randomly saw my mother coming out of a dentist appointment in the building directly across the street from the parking lot. As we spoke, I could tell she was in shock just as much as I was. It was at that moment, around 10:30 a.m., that my mother, a native New Yorker, told me in tears that the terrorists who hijacked the planes and crashed them into the World Trade Center towers had managed to bring both buildings to the ground. I was shocked beyond belief at hearing this information. My next question was if my sister had been accounted for, and had my mother spoken to her. I was grateful to hear my mother say she did speak to her just minutes before, and that she was at home in New York. Although she was emotionally shaken up at the time, as all New York citizens were, she was all right and accounted for. As grateful as I was that she was okay, I then became sick to my stomach thinking of all the thousands of ordinary people like me who were not as fortunate and were savagely killed and wounded by simply traveling, or going to work, as I had done that day.

I was in complete shock, and I waited for my father to come pick up my mother at the parking lot. When he did, I told them I would call them later. I went over to a friend's house, unsure how to grieve or what to do. I just knew that I did not want to go home and watch this horrific tragedy on television by myself. Although my friend had not yet come home from work, his mother was there, and we watched the events of horror over and over on television. They did confirm on T.V. that the earlier suspicion of ten more planes was possible; they were just unaccounted for at the time, but it was unknown whether they were hijacked or not. Ironically enough, this was the same house in which I watched the events of the attempted

11

assassination of President Reagan twenty years earlier. I knew that this attack on the United States was going to change how we lived for the foreseeable future. When my friend did come home, we continued watching the events in disbelief, and our sorrow turned to anger over how an event as tragic as this could be allowed to take place in our country. This tragedy showed that, in this new century, we as citizens and our soldiers fighting overseas in enemy territory are involved in a new type of warfare, where no one is safe from the enemy.

From the hijackings of September 11th, 2001, and the anthrax attacks that followed, the comfort that we once felt in this country was tarnished forever by those evil ones whose only intent is to make innocent people suffer while spreading their message of hate. Has the world become so evil that people who pervert religion are willing to kill as many innocent people as possible just to advance their own ideology? Are they willing to inflict as much destruction as possible to end our country as we know it? Have these terrorists become the newest enemy not only to America, but also to the Western world? Is our military, still the finest and most powerful force in the world, up to the task of wiping out the new enemy and working closely with our federal government agencies to prevent such atrocities from ever occurring again? Will the United States be attacked again in the manner of 9/11, or to an even worse degree in the future?

Why do I make these points to start the book? I believe that we are all profoundly affected by the human tragedy that occurred on that day. It is hard to do one's daily activities and enjoy life the way it was before 9/11, knowing that we as individuals are no longer safe from the profound evil that clearly exists in the hostile world we live in. Obvious questions that we can ask ourselves and each other are: How can we feel safe again? Has this nation ever been involved in conflicts against terrorism before? Have we ever been in a war situation in which we were on the verge of losing our liberty, freedom, and very lives? Can we overcome the tragedy issued to us on 9/11 to become safer from terrorists, and more prosperous? Can we win the war on terror? Can we ever feel safe from this type of evil again?

In this book, I believe that this great country has been through similar situations of national calamity and not only survived, but grew to prosper economically, socially, and militarily. In order to adapt and learn from our current situation, history would dictate that it has lessons to be examined from our past struggles. From the Revolutionary conflict with the British in the latter half of the 18th century to the current war on terror, the opponent has always remained intense and formidable. What can be learned from the past wars that can be beneficial to our present and future populations so that terrorism can be prevented against civilian populations in our country and other countries around the world?

In this book, I examine the past presidencies of those who called our nation to war, and the Presidents who would arise sometimes as a result of their military performances in times of conflict. I will also examine the status of the current war on terror that we are now in under the leadership of former President George W. Bush, former President Barack Obama, and President Donald Trump. With the focus on past and present war-time Presidents, I show how America has conquered past foreign adversaries with presidential leadership that was capable of adapting to the changing climates of particular war-time scenarios that each President faced during their time in office. In doing this project, I hope that by examining past presidential actions in war conflicts, we can compare the solutions that our past leaders reached to our own solutions that we as a nation can build on to strengthen and secure our nation. For it is true that this nation has been through similar types of dogmatic and unprovoked attacks. Throughout our entire history, the United States has been forced to make the ultimate decision of sacrifice human lives in war to benefit our nation and free people worldwide. It is also true that our country has fought military conflicts against other nations for different reasons, some of which were not always as popular with the public as others have been.

At the end of each chapter, I want to list the improvements that each Executive involved in war brought to the United States, and point out that some conflicts may not have been as successful as others. I wish to prove how these individual Presidents, along with their partners in Congress and leaders in the military, made us obtain greater strength in the ability to improve our citizens' lives and improve America's stature on the world stage. In my analysis of all of America's wars in history, some were fought for the "right" reasons, but the public may have disagreed at the time and may have considered the reasoning to be ambiguous or controversial.

Despite the senseless and morally corrupt tragedy that took place on our soil on September 11th, 2001, I will prove that our current enemy can and will be defeated through resilience and determination from our federal government. I will also prove to those who tragically lost loved ones in the attacks of that day, and to all those who were affected by it, that, as President Lincoln said so brilliantly during his Gettysburg Address in 1863: "They will not have died in vain, that this nation, under God, shall have a new birth of freedom, and that government shall not perish from the earth." Those words ring true now in our world more than ever.

Chapter 2
The American Revolution (1775-83)

The Revolutionary War that took place from 1775 to 1782 was the conflict that granted America's freedom and independence from England. The thirteen colonies actually merged together under one federal unit to defeat the more powerful England on their home soil, familiar geographic terrain to the American army. While the English army had help in this war from the German mercenary troops (Hessians), Native American tribes, and Loyalists, the Colonists had crucial assistance in their war campaign from France, Spain, and Poland.

There were many reasons for this war, but perhaps the biggest reason was the extremely heavy taxation that was being levied against the colonists by the British, due to the debt amassed by England from the Seven Years' War (French-Indian) with France. This figure was estimated to be 140 million pounds (pg 40, #5). This fact did not sit well with Americans, because the colonists had no representation in Parliament, and the taxes levied against Americans were simply too unjust. This is where the phrase, "Taxation without representation is Tyranny!" came from (pg 8, #8). Due to these feelings of oppression that Americans felt from their mother country, they felt that they had no choice but to revolt, or get much more freedom from England.

The colonists believed that the distance between the colonies and England was too far to have fair, effective representation in Parliament. Another reason for the American coalition was because England was growing more oppressive towards them. From England's point of view, the colonies existed solely for the enrichment of the mother country, and were there to provide England with gold, silver, raw materials, and markets. England had dominant control over America because of the victory that Britain achieved in the Seven Years' War from 1756 to 1763. This victory for England forced the French control over certain areas in North America to be surrendered to England, which included the entire St. Lawrence Valley and the territory between the Appalachian Mountains and the Mississippi River. This acquisition in America made England the most powerful country on Earth at the time.

This fact would eventually cause controversy between the two parties of England and America. The results from this war between England and

France in America caused King George the III and his English government to go into massive financial debt, which they thought could be solved by heavy taxation of the American colonists, because they had no representation. This heavy taxation began when England enforced the Sugar Act of 1765, which heavily taxed the sugar being brought to the colonists (pg 68, #8).

King George the III was only 22 years old when he claimed the English throne in 1760 from his dead grandfather, who had produced no male heirs.(pg 32, #5). By the time of the problems between England and the colonies, the King was in his mid twenties. This was not considered unusual, because most members of Parliament at the time were simply the sons of the fathers who preceded them, and sometimes these men inherited power in their early to mid twenties (pg 33, #8).

At the time, people who did not own property in England had no representation in Parliament. The British monarchy simply decreed this fact as "God's will." As the costs of the Seven Years' War to England were estimated at 140 million pounds, the easy choice to solve this financial crisis was to tax the American colonists. However, colonial representatives in America, led by Patrick Henry, complained of this unfair tax directly to the English authorities in government, and successfully persuaded England to withdraw the tax.

Unfortunately, England did pass the Townshend Act, named for Chancellor Charles Townshend. This act imposed duties on tea, lead, glass, paper, and paints that entered America's harbors. England also put barriers on settlers and ordered them not to cross the Appalachian divide (pg 97, #1). It became increasingly clear to the colonists that the colonies were being used as trade weapons by England to compete with France, Spain, and Holland (pg 22, #8). Adding to the discrimination was the fact that British judges were known to be unfair and biased against the American colonists, and were concerned solely with the interests of the Crown (pg 24,-#8).

These acts of hostility directed at American colonists began to exacerbate the situation, and angered some American leaders, including George Mason. Even George Washington, who had served England in the French-Indian War, found the unfair taxes being levied against Americans disagreeable. As a result, American leaders met unofficially with most members of the House of Burgesses in Williamsburg to establish a boycott of British goods. This tactic was successful because England did, consequently, repeal all taxes on British revenue in 1769, with one exception: the tax on tea. The British East India Company benefited greatly from the law, and consequently reaped substantial profits as a result. This act by England drove the colonists to begin the course of events that would become the Revolutionary War.

During this time, George Washington was serving as the Justice of the Peace of Fairfax County, Virginia. Washington had this role during and after the French-Indian War, from 1760 to 1774 (pg 123,-#8). In order to understand why the colonists chose George Washington to lead them into war against a more powerful English military and, after achieving victory over England, to become our nation's first President by popular choice, I believe that it is important to understand Washington's entire history, which guided him on his path to power in a unique age of democratic birth.

Washington was born on February 11th, 1732 in Westmoreland County, Virginia (pg 3, #13). However, due to a defect in the calendar that existed in Great Britain at the time, eleven days were added to Washington's birthday, which made it February 22nd. Washington was not born into extreme wealth and prestige, nor was he a member of the first families of Virginia, nor a member of the Imperial Purple. In fact, there was little to suggest that this American subject of the British Crown would play an important role in this part of the world that Europe seemed to dominate.

When Washington was a young man, he inherited property called Ferry Farm, and began to work as a land surveyor of Culpeper County, Virginia. This was located in Rappahannock, Virginia. Washington's brother, Lawrence, inherited Mt. Vernon from his wife, who was the daughter of Colonel William Fairfax. The fact that Fairfax County in northern Virginia is directly next to the city of Washington, D.C. is certainly no coincidence, Washington's work as a surveyor was perfectly suited to his passion for nature. In 1752, his brother, Lawrence, fell ill and died, leaving George without an important mentor in his life. As a result of Lawrence's death, George inherited Mt. Vernon.

The reality of Washington's world, at the time, was very harsh. Life was full of diseases, sickness, or Native American attacks, and death could come in multiple ways. While George was looking for a suitable wife to spend her life with him at Mt. Vernon, he fell in love with a married woman named Sally Fairfax, with whom his family had friendly connections. However, this romance was not pursued by Washington, because had he done so, it could have led to a duel, which was a popular way of settling disputes at the time.

As for the aforementioned French-Indian War that occurred from 1756 to 1763, Washington would gain real military experience as a member of Britain's campaign against France. The French already had settlements in North America, which included Quebec and New Orleans. These French settlements had formed alliances with native tribes that attacked the colonists, and drove the settlers and traders from the Ohio and Mississippi River areas. Colonel Washington was given the dangerous mission of entering the French-controlled territory that was located just across the Alleghenies, to demand to the French army that it leave British territory. Once Washington was given the French response rejecting these terms, he returned to his men and began to prepare his troops for battle.

Washington and his men fought bravely at the Battle of Ft. Necessity, but would lose to the French-Indian military combination, which was led by French military leader Coulon de Villers. His men would outnumber Washington's army by 900 to 400. This experience of defeat at Ft. Necessity was of great importance in forming the character of Washington and enhancing his leadership skills. Controversy did develop when Washington accepted the terms of defeat from the French, who had gotten back to English military leaders. In the terms of the battle loss, Washington agreed to sign a document to the French commanders that stated that Washington and his army committed murder against their soldiers and also admitted responsibility for the English soldiers' fatalities.

Despite this controversy, Washington's explanation of this event was that, because he was not fluent in French and he was forced to rely on imprecise information from his translator at the time, he had no other choice but to sign the treaty. This act would save the lives of his remaining regiment and allow him to re-group with his fighting men in another location (pg 30,-#1). However, due to the controversy that this event had created in Britain, Washington actually lost his rank of Colonel and briefly retired from his military career in 1754. Washington went back to his home at Mt. Vernon, Virginia.

After only a short hiatus there, Washington returned as a volunteer adviser for General Edward Braddock. This is another example of Washington's bravery and courage, because he once again entered into battle for the British, and therefore risked his life for victory. Because of his bravery, Washington was restored to a high rank in Britain's military service. For example, Washington displayed tremendous courage in his participation during the Battle of the Monongahela River, in which two thirds of Braddock's forces, including Braddock himself, were killed by a musket ball. Other members of Braddock's forces were captured and tortured, some to death, by their French-Indian captors. During this point of the crisis, Washington stepped into military command and led his troops to safety.

As a result, Washington was given true hero status by the troops, who were proud to serve with him. This also included his past bravery in earlier military campaigns. Washington's popularity soared back in England as well, especially in the House of Burgesses. Washington was becoming a true celebrity in the colonies at a mere 22 years of age. Washington was immediately reinstated in Great Britain's army at the rank of Colonel, and was placed in charge of all Virginia forces (pg 45, #1). Washington's experiences from the past, which included the controversial surrender bargain with the French and also the pain of witnessing many of the troops he commanded mercilessly slaughtered in battle, redefined his character and strengthened his abilities to be a successful military commander.

Washington would adapt his command ability to be more precise and have stronger discipline. For example, he would not allow his soldiers to remain subjects of drunkenness and profanity, for those who would continue such behavior would be punished (pg 48, #1). Washington saw and learned during battle against the French-Indian alliance that the specifically distinct red colors of the British uniform gave them a disadvantage in fighting. Therefore, Washington was responsible for the change in dress of his officers from the distinct red to the blue cloth material (pg 48,-#1). In order to keep his men's spirits as high as he could, Washington was able to successfully negotiate a pay raise for his fighting men. Washington's negotiating skills were exemplary in also convincing the Cherokee and Catawba tribes of Native Americans to aid his forces during his military offenses.

During the later stages of the Seven Years' War campaign, Washington's physical health suffered from bouts with dysentery, which greatly affected him. This sickness lasted for the remainder of his service in the French-Indian conflict, as well as future conflicts against England in the years to come. The war continued with some stunning triumphs for Britain, including the capture of the city of Quebec and some parts of the country of Canada (pg 71, #1). These victories led to Britain's triumph in the war versus France that ended in 1763.

Upon Washington's return home from the war, and after a failed courtship with the previously mentioned Sally Fairfax, his romantic attention was diverted to a wealthy woman named Martha Custis, who was a widow from her marriage to Daniel Custis. Custis was a rich planter in Virginia, and upon his death, Martha had inherited his fortune. Martha was the mother of two children, Jack and Patsy. When Washington married Martha, he inherited 7,000 acres of land. Washington's marriage to Martha also brought him the status and high credibility that he was searching for in securing his status among the fabled First Families of Virginia. This was an aristocratic group that Washington wanted to belong to.

Washington also inherited the role as stepfather to Martha's two children, and Washington gave all his efforts to fulfill his role as a loving father to them. Washington stressed the importance of education to his stepchildren, and he made sure that they had the best of all resources at their disposal for their benefit. Unfortunately and tragically for Washington and his family, his stepdaughter Patsy developed a case of epilepsy, for which there was no cure at the time. Despite the attempts that George and Martha made at curing Patsy, which included such medical practices as bleeding and panaceas, she died on June 19th, 1773. This tragedy would remain with Washington for the rest of his life.

In spite of the sadness of Patsy's death, Washington was determined to go forward with his life of being a farmer and making successful use of his land and crops. In the beginning of Washington's return to farming, his tobacco crop was not as profitable as it should have been and, therefore, he was in debt to England for the services rendered that would provide relief for his unprofitable product. Because he owed a lot of money to England, Washington altered his cultivation to include wheat on his farm. Washington continued this practice by cultivating other products on his farm, such as oats, corn, alfalfa, timothy, clover, flax, hemp, peaches, apples, and cherries. Washington was successful in merging his skills as a land surveyor and a farmer to turn his farm around into an economic success.

Because of the land that Washington obtained from British General Robert Dinwiddie from his service at the conclusion of the Seven Years' War, he was able to grow his land ownership from 7,000 acres to 30,000 acres, most of which were located beyond the Appalachian Mountains. This made Washington the most successful land speculator of all time (pg 93, #1). Due to his rising reputation, Washington became a natural choice as a leader for the colonies when trouble with Great Britain began to arise.

Washington saw that the rights of the colonists were being abused by England, who used the colonists as trade weapons to compete with France, Spain, and Holland (pg 22, #5.). Washington's allegiance was firmly with the colonists. After the protests mounted from the colonists for being unfairly taxed with no representation in Parliament, an incident occurred on March 5th, 1770, in which five American protesters were killed in a clash with British soldiers. Ironically, future President John Adams and Josiah Quincy would legally defend the British troops. At the end of the trial, four of the six soldiers were acquitted, and the remaining two were found guilty of a misdemeanor crime and subsequently discharged from the military. This act would bring outrage from America, but it failed to bring about immediate rebellion (pg 87 #8).

That was, until June 1772, when a group of colonists captured the hated British vessel *Gaspee* off the coast of Rhode Island. The crew of the *Gaspee* were taken prisoner, and the ship was set on fire. This act was followed by the American response to Britain's oppression and unfair taxes at Boston Harbor on December 16th, 1773, when civilians, dressed as Indians, poured the English tea into Boston Harbor to protest the unfair taxes being levied against them. As a result, King George the III closed Boston as a port, and General Thomas Gage was sent from England to take control of the Massachusetts Colony (pg 92 #8).

When the First Continental Congress met on September 4th, 1774, war between England and the colonists seemed inevitable. The men who represented the seven colonies that were present were Richard Henry Lee, Patrick Henry, Richard Bland, Benjamin Harrison, Edmund Pendleton, and George Washington. Washington was chosen to lead the Revolutionary Force as general. Washington stood firmly behind the rebels against England (pg 104 #8). The war for independence was now at hand!

The Battles of Lexington and Concord, Massachusetts began the Revolutionary War on April 19th, 1775. England already had a garrison of 4,000 troops in Boston at the time. In the Lexington battle, the American military sustained eight fatalities and ten wounded, compared to one British soldier wounded. However, the statistics would be different in the Battle of Concord, which resulted in 73 British troops killed, 174 wounded, and 26 missing, while the colonists suffered 49 soldiers killed, 41 additional wounded, and five missing.

The psychological effect of the Battles of Lexington and Concord was to unite the colonists to rebel against England (pg 115 #8). The colonies wanted to appear united because England wanted them to act as individuals, with no cohesion (pg 116- #8). John Adams believed in unity, as did Washington, who wanted a Continental Army from members of all colonies (pg 121- #8). By the end of 1775, Congress had successfully added 27,500 soldiers to its payroll, which is amazing considering that there was no army whatsoever in the previous year of 1774 (pg 122 #8).

The American generals who served in the war were Artemas Ward of Massachusetts, Phillip Schuyler, Richard Montgomery, Horatio Gates, Charles Lee, and George Washington. For England, Generals Henry Clinton, John Burgoyne, and William Howe served in this war. Washington inherited 14,500 men to train and discipline when he was promoted to General as Military Commander for the colonies (pg 125 #5). At the time, Congress was the American government.

With England determined to crush the rebellion, they employed the Hessians from the then German states to help destroy the Americans. The British also employed some Native American tribes to help them fight the colonists, and they promised the Native Americans land rewards if England were to be victorious. America wanted to enlist the services of either France or Spain to help them out, because these countries had conflicts with England, too, at the time. Britain had a strong navy, and ready capital at their possession, which could easily produce war material. However, the war with America would prove to be very unpopular in England.

By January 1st, 1776, Washington's army dwindled to 10,500 men (pg 131 #8). However, by July 4th, 1776, the Colonists rallied together and signed the official Constitution of the United States in Philadelphia, mostly written by Thomas Jefferson and James Madison (pg 140-#8). After the battles of Lexington and Concord, the British concentrated their attacks in 1776 at New York, and overwhelmed American forces there, because the British wanted control of the Hudson River. This defeat by the British made Washington and his forces flee across the Delaware River into Pennsylvania. However, when Washington evacuated New York City, knowing that the British were about to occupy it, he and his army burned most of it to the ground, so that it would serve little use to the

English. Washington did successfully move his troops out of harm's way in Manhattan (pg 269 #8).

Washington would also receive defeat at Ft. Lee, and was forced to retreat to New Jersey, with British General Cornwallis in hot pursuit. Washington's problem, at the time, was two-fold:

1.   Form and train an army on the battlefield
2.   Raise and recruit a new one to replace the old (pg 179 #8).

Meanwhile, General Lee was captured by the British on December 14th, 1776, in New York (pg 154- #8). While on the run, Washington and his forces counter-attacked the British at Trenton and Bordertown, New Jersey on Christmas night, 1776. These attacks were successful for American forces. In fact, Washington successfully captured 920 Hessians during the Trenton surprise attack. As a result, Washington was promoted again for his accomplishments. This victory also brought in new, important recruits for the American army.

France and Spain were secretly sending supplies to America, such as clothing, artillery, and cash. This occurred because both these nations wanted to see the country that humiliated them in the Seven Years' War beaten by their American counterparts. Yet, America could not convince Canada to unite with it to defeat England. This resulted in the unfortunate defeat of U.S. troops in December 1775 at the Battle of Quebec, and the loss of U.S. military control over Lake Champlain. (pg 144, #8). Despite these losses, Americans strongly believed in the cause they were fighting for. Plus, Americans were fighting at home and knew the land terrain better then England did.

Yet, when England landed in New York on August 26th, 1776, with over 31,000 troops, America's 19,000 troops were overwhelmed, and Washington lost 970 men who were killed, wounded, or missing, compared to Howe's 63 killed and 337 missing or wounded (pg 264, #5). However, Washington succeeded in evacuating his troops to New Jersey, and saved the mission. Washington knew that he could occasionally lose cities or battles, but he would never give up on American independence, even after losing Forts Lee and Washington.

Washington and his troops would come back in the war by winning the battles of Trenton and Princeton, New Jersey, but they were still forced to retreat, because the Patriots would have been badly outnumbered if England counter-attacked (pg 192, #8). By this time, France's King Louis the 16th started to send fresh supplies to the American troops, which were badly needed (pg 195, #8). In fact, Benjamin Franklin was sent to Versailles on behalf of the Continental Congress to request further financial and military assistance from Louis the 16th on December 28th, 1776. France, however, wanted to wait on this proposal until there were more U.S. military victories (#11).

This diplomacy that Washington demonstrated was winning morale among his troops, and was convincing them to re-enlist in order to gain

21

momentum against the British military. In May 1777, Washington was in control of 7,000 troops at the Morristown, New Jersey camp, with other troops also stationed at Ft. Ticonderoga to oppose a British attack from Canada. A smaller force was in Peekskill to delay any British movements up the Hudson River.

However, the British army would instead move south towards the Chesapeake Bay in an attempt to mount an aggressive offensive military campaign against Philadelphia. Meanwhile, the British successfully recaptured Ft. Ticonderoga on July 5th, 1777, led by British General John Burgoyne: 2,500 American troops were overmatched by 7,000 British troops. Burgoyne wanted to control the Hudson River to cut off American supplies that were traveling on it. However, Washington had decided to remove the captured weapons from Ft. Ticonderoga and move them for use against the British. The English army employed the Hessian troops from the then-German states (pg 206, #8). Washington took 15,000 American troops to Philadelphia in order to fortify American defenses against the impending British attack. But the city of Philadelphia fell to the British and Hessian forces anyway, after they had successfully invaded and beaten the American army on September 11th, 1777. British forces were led by General William Howe. This was the Battle of Brandywine Creek.

The British also successfully won the Battle of Brandywine, New Jersey, which forced Washington and his troops to retreat to Chester, Pennsylvania (pg 213, #8). Then, the Paoli Massacre took place on September 20th-21st, 1777, when the British surprised unsuspecting troops and bayoneted approximately 150 men. As bad as this situation seemed for the colonists, not only had the Continental Army escaped Philadelphia, but Congress also escaped Philadelphia and were operating in York, Pennsylvania (pg 216, #8). The Americans were constantly slipping away from British forces, such as when the Americans abandoned Manhattan in New York City and burned most of the town as they left, so that the British could not have it for their own purposes. Even though Washington and his men were on constant retreat from British forces, Washington was always planning for an opportunity to strike the enemy.

Washington personally led the attacks against the British at Germantown and Princeton and against the Hessians in Trenton. Although the Americans technically did not win these battles, they showed themselves, and the world, that they could legitimately fight a more powerful England and win if they just stayed dedicated to the cause. More importantly, despite Washington and his army losing the invasion of Germantown, it did bring the country of France back into the war, because the French discovered Washington's fine leadership skills (pg 218, #8).

On October 7th, 1777, the Battle of Saratoga took place, which ended in a decisive American victory (pg 226 -#8). Horatio Gates was the American general in this region of the Northeast. Gates and Burgoyne were once allies in the French-Indian War. However, they met again as enemies at Saratoga. The two big battles of Saratoga were Freeman's Farm and Bemis Heights. The Battle of Saratoga turned the tide of the American Revolution campaign against Britain, and then the French alliance with America began (pg 417- #5). General Burgoyne was commanding British forces in this battle, and he was cut off from help from General Howe, who happened to be stuck in Philadelphia.

The British were repelled and Gates was granted the victory. On October 17th, 1777, Burgoyne surrendered 6,000 British soldiers as well as his own sword. Howe did not help Burgoyne. Because of Burgoyne's surrender at Saratoga, he was sent back to England, and he was never given command of a British army again (pg 227, #5). However, both Delaware forts that were constructed by Americans, the Mifflin and the Mercer, fell to the British (pg 239, #5). Because of these victories in battle, and the fact that the course of the war was coming upon the winter season, the British decided to rest in comfort within the city of Philadelphia, while the rebels suffered extremely in the cold at Valley Forge, Pennsylvania.

Washington chose Valley Forge to retreat to in the winter of 1777-78 because it was centrally located and easy to defend. He could also watch the British, who were 23 miles away in Philadelphia. In the meantime, Ben Franklin, Silas Deane, and Arthur Lee worked out a formal treaty with France in return for their cooperation with America in fighting England (pg 243, #5). Back at Valley Forge, the rebels did not even have shoes or clothing to battle the bitter cold. Having supplies on hand was a huge problem for American forces, and Washington's army was literally starving. Witnessing his men suffer, Washington said, "The injuries we have received from the British nation were so unprovoked, and have been so great and so many, that they can not be forgotten" (pg 169, #1). By February 1778, 2,500 American troops would perish from disease.

Washington's opponent, General Howe, was content, but not happy, with occupying Philadelphia, because the city was of little use to England. This was because all of the important people and documents in the city had already been evacuated. By that point, England had already won battles at Brandywine, Germantown, Skensboro, Ticonderoga, and Forts George and Edward. When the British decided to evacuate Philadelphia, Washington led his men to a successful recapture of the city.

In February 1778, General Howe was decommissioned by England and was sent back home. Howe was replaced by Sir Henry Clinton (pg 245, #8). When Washington was back in Philadelphia, he saw the economic problems that inflation was creating within his armed services, and also noticed the profits that merchants and speculators were making. These decisions and experience with inflation and the economy would serve him well in his later years as President.

In early 1778, a Prussian general named Ferdinand von Steuben arrived in Valley Forge to help train American troops to perform better in combat. Also, when French forces arrived at Valley Forge, especially Marquis de Lafayette, they were very impressed with the Americans' will to fight England, even though they lived in miserable conditions. Consequently, the French army supplied and trained Washington's army, which gave significant strength to America. By May 5th, 1778, 13,000 American troops were strong and well trained (#11). After France and America had allied themselves in early 1778, the British chose to attack the southern United States, because it contained more valuable crops than the north had (pg 273, #5). Robert E. Lee's grandfather, "Lighthorse Harry" Lee, served brilliantly for the colonists. Nathanael Greene was also an American general in charge of troops, at just 33 years old. He helped rebuild supply lines for the Americans.

England wanted to invade the southern colonies because the southern colonists had more loyalty to England than the northern ones (pg 269, #5). In fact, it is estimated that as much as a third of the American population of the time were loyalists to England, and they were called Tories (pg 271, #5). Meanwhile, back in England, an American captain named John Paul Jones wanted to attack the country of England itself from the sea. Jones did successfully attack supply ships and sea ports in order to hurt England's more powerful navy. The ship *Bonhomme Richard* was later commanded by Jones.

By the time Henry Clinton took over for General Howe as British commander, England still had control over Philadelphia, New York, and Newport, Rhode Island. The French, however, could blockade any of these ports. Accordingly, the English government told Clinton to evacuate Philadelphia in order to protect other ports. The British treated the city of Philadelphia terribly during its occupancy. Meanwhile, Washington and his forces would lose to the British at Germantown.

Washington gave command to General Charles Lee to attack the evacuating British soldiers coming out of Philadelphia. Lee fought in the French-Indian War, had been known for his arrogance, and was deemed by others to be suspicious. On June 27th, 1778, at Mount Holly, New Jersey, Washington gave the order to Lee to attack, and 13,000 American troops began their assignment to go after the English. This occurred at a site called the Monmouth Courthouse. At the beginning of the battle, Lee hesitated and did not attack as he had promised to do, so Washington relieved Lee of his command. Washington himself then took over as commander. On a 100-degree day. Washington rallied the American troops in Lee's absence, and America had a good defensive position. Over 20,000 troops would clash on that day at Monmouth. Intense combat lasted for over five hours. Soldiers died from the heat as well. Washington's troops performed well, but the battle was a technical draw.

Lee was later court-martialed for his actions, and he died in 1782. Philadelphia was now back in American control.

Washington slept on the battlefield with his soldiers. The Continental Congress was back in session in Philadelphia. Meanwhile, John Adams had shown up in France to assist Ben Franklin's quest to obtain French help. They were successful in acquiring the French navy for help, which was badly needed. Another problem for American soldiers was that, by the fall of 1778, they were not being paid for their services. Slaves were being recruited for America; in return, they would be granted their freedom for their military service. These slaves went to Rhode Island and formed the 1st Rhode Island regiment, and they helped the Patriots' cause tremendously.

The British forces had allies among the Native American tribes, including the Shawnee, Iroquois, Delaware, Ottawa, Cherokee, Wyandots, and Mingos, who sided with them against America. The Kentucky region, during the Revolutionary War, involved British and Native American forces teaming up to fight America. The Native Americans also controlled most of the upper Northwest. However, two tribes of Native Americans did not side with the British. England did capture Savannah, Georgia, on November 27th, 1778, and occupied it until July 1782. Yet, the city of Charleston was still in Patriot hands, at least for the time being.

Meanwhile, Washington remained vigilant in his attacks on the Iroquois, whose allegiance lay with the Loyalists. In June 1778, 500 Iroquois warriors attacked the Wyoming Valley in northern Pennsylvania and killed 200 frontiersmen. Washington responded to these attacks by giving the order to invade the Iroquois area from southwest Pittsburgh, from contingencies led by Colonels John Brodland and John Sullivan. Together, these forces inflicted a large blow to the Six Nation Alliance of the Native Americans, from which they never fully recovered. (This included the Shawnee tribe and the Cherokee nation.) In August 1779, 4,500 American soldiers attacked Newtown, New York, and burned the village to the ground. The Native Americans would relocate to Niagara.

However, back in Georgia, Britain won some significant battles against American General Benjamin Lincoln in Charleston, South Carolina, and successfully captured Lincoln's 5,500 troops and all of their supplies. This was an important loss for America because of Charleston's port (pg 278, #8). It is said by some historians that the fall of Charleston was the worst Patriot loss of the entire war. South Carolina and Georgia would eventually surrender to Clinton at the end of 1778. As a result, thousands of Yankee P.O.W.s were brutally murdered by the British.

Meanwhile, American military leader Benedict Arnold was angry with American authorities who had allied themselves with France, because he hated monarchies and Catholics (pg 281, #5). Arnold was appointed Philadelphia's military governor, but in truth he did not want the Philadelphia job. Arnold battled ferociously with Gates over issues of authority. Arnold was always at odds with American military leadership. Eventually, Arnold was accused of abuse of power and had to face his charges in court on March 5th, 1779. Accordingly, Arnold called British Major John André to sell out the American army for money. Arnold actually had plotted with England to surrender West Point, where he was stationed as Commander, to the English authorities. However, when Arnold's plan was unraveled by American authorities, he successfully fled to England as a deserter, and his co-conspirator, Major John André, was hung for his role with Arnold. Arnold's defection to England was a surprise to both George Washington and to his army, because Arnold was once considered a brave and courageous leader. In fact, Washington was very stunned by Arnold's betrayal. England offered Arnold 20,000 pounds and military command for England, and he took it.

Another problem for America was keeping its own soldiers from mutinying, because of the irregular pay they received as well as shortages of food and clothing (pg 288, #8). However, the good news for America was that by June 16th, 1779, Spain had joined the fight with America against England (pg 445, #5). Washington decided to appoint Nathanael Greene to lead the southern war theater against General Cornwallis. General Cornwallis invaded Virginia as far as Charlottesville, which forced Governor Thomas Jefferson to flee Monticello.

In December 1779, the war was at a stalemate. Neither side had an advantage. Henry Clinton controlled New York, but Washington was nearby. Morristown, New Jersey was the American winter camp that year. It was a very brutally cold winter again. This was a low point in Washington's career: he received no help from the Continental Congress, and U.S. currency was practically worthless. However, English national debt was very high, and the war in America was very controversial. The reasons for this unpopularity were that it was financially very costly and that France had sided with America.

Clinton received orders on December 26th, 1779 to go on the offensive against the colonists in the south. A lot of pressure was placed on Clinton. The British left New York with a third of their army to head to Charleston, South Carolina. Washington could not send supporting American troops to help the American southern army, led by Major General Benjamin Lincoln. There were 2,400 soldiers under Lincoln's command. By January 1780, 8,700 British soldiers were on their way by ship to South Carolina.

Lincoln was from New England. Charleston, South Carolina was surrounded by bodies of water. By January 10th, 1780, Lincoln was told of the British forces that were on their way. Congress promised Lincoln an additional 3,000 troops, but he still faced terrible odds. On February 11th, 1780, Clinton and his soldiers landed 20 miles south of Charleston. Clinton did not want to destroy the town, but rather wished for a forced surrender. The British offered the slaves freedom if they would fight for the English. By April 1st, 1780, the siege began. After a week, American reinforcements showed up to Charleston, but with only 750 men. Ten thousand British troops were already there; Lincoln was in huge trouble. Britain fired mortars into the town every day. After 41 days of battle, Britain was the decisive winner. America surrendered on May 12th, 1780, and 5,000 prisoners were taken by the British. This was the heaviest loss of the Revolutionary War. Charleston was officially captured by the British, and Lincoln officially surrendered his sword.

A Loyalist militia was trained in the south to fight American forces. Loyalist militias attacked American citizens in the back country in June 1780. Meanwhile, Lord Banastre Tarleton took over for England in the south. Tarleton was cruel and brutal, and he was known for attacking soldiers already in surrender mode. Such was the case at the Waxhaws battle. During this time, General Cornwallis was appointed to replace Clinton to lead the English southern army. Cornwallis' wife had just died in England, and after her burial, he had returned to America. Congress chose Horatio Gates to lead the Continental southern army, over Washington's severe objections. Gates was the hero of Saratoga three years earlier, but Gates' army was starving. The battle of Camden, South Carolina was on August 16th, 1780, with 5,000 English troops battling 3,000 American soldiers. Lord Tarleton led the British side. Gates would be beaten in battle by Tarleton, and Gates was subsequently court-martialed. Washington proved to be right about Gates. Consequently, Washington picked Nathanael Greene to replace him.

Because English forces controlled the Chesapeake, they had built their headquarters at Yorktown, Virginia, which was located between the James and York Rivers. In response, Washington ordered his troops to New York, so that General Clinton could be stalled from helping the re-instated Cornwallis in Virginia. Even though Cornwallis had all but captured South Carolina in 1780, the British were defeated at King's Mountain, North Carolina. Also, on January 16th, 1781, Daniel Morgan led America into battle at Cowpens, South Carolina. Morgan set a trap for Tarleton, and it worked. America would win that battle. This defeat sent Cornwallis back to South Carolina. The British had technically won the Battle of Guilford Courthouse in 1780, but on March 15th, 1781, Cornwallis fired a cannon into his own men in order to kill Americans, and was forced to retreat to Virginia. Even though more British died at this battle than did Americans, England still technically won with a controversial victory. This would set the stage for Yorktown (#7).

By September 1780, Henry Clinton was in New York City. The British and French were fighting each other all around the world. Generals Rochambeau and Lafayette met with Washington to discuss a formal alliance. The newly arrived General Rochambeau of France was originally scheduled to invade New York by General Washington's order. However, the French navy sought to corner the English at Yorktown, despite the fact that Washington wanted to strike New York. After some convincing by the

French, and sensing that New York was too heavily fortified by the British, and because he had learned the news that the French navy was on its way to help America, General Washington ordered a complete attack on the city of Yorktown, Virginia, with support from Generals Lafayette and DeGrasse (pg 321, #8). The plan was to have General Clinton fooled into thinking that Washington and his forces were coming to New York. This took place in 1781.

General Clinton had to stay in New York, because he assumed that Patriot forces were coming there, and he also assumed that Cornwallis could successfully repel Patriot forces coming towards Virginia. Unfortunately for England, due to overwhelming French naval forces arriving off the Virginia coast, the Patriots successfully inflicted heavy damage on British ships. The plot was to surround Cornwallis at Yorktown and force him to surrender. The American army had Cornwallis surrounded, and the French navy had England cut off in the Chesapeake. General Cornwallis was not only cut off from supporting British troops by land, but also by sea.

General Washington led a Patriot attack of 8,800 American troops and 7,800 French troops into battle at Yorktown. General Cornwallis, who was in dire straits, wrote to General Clinton, "If you can not relieve me very soon, you must be prepared to learn the worst!" Colonel Alexander Hamilton led his forces into Yorktown, whose manpower both outnumbered and outgunned the British. This battle occurred from October 15th-17th, 1781. As correctly predicted by Washington, Clinton could not give Cornwallis the military support that he needed. Consequently, Howe and his British forces surrendered to the Americans on October 19th, 1781.

As Cornwallis sent another general, Charles O'Hara, to surrender the sword, Washington sent Benjamin Lincoln to accept it. This was because of Lincoln's humiliating defeat at Charleston, South Carolina, to the British. Washington realized the significance of Lincoln having the pleasure of appearing as the victor. England, forced to realize that they had doubled their national debt and that the war in America was very unpopular at home, had to formally acknowledge American independence. The war was over, and America had won. This victory over England was considered to be a huge upset.

The trouble between the Native Americans and white settlers would continue, especially in the Northwest Province, all the way until the War of 1812. The total costs of the Revolutionary War to the United States were as follows: in the eight years of war (technically, the Treaty of Paris was signed in 1783), 231,771 soldiers served for the United States, but no more than 35,000 at any one time; 4,435 American soldiers died in battle, and 6,188 were wounded. The English numbers are approximately the same (pg 335, #8). When King George the III finally accepted defeat, a new Prime Minister, named Lord Shelburne, favored American independence. Remaining English troops had to evacuate the states, which they did. British troops had disappeared from New York City by November 1783.

The British left their two southern strongholds of Savannah, Georgia and Charleston on July 11th, 1782 and December 14th, 1782, respectively. The Treaty of Paris was signed on September 3rd, 1783, which granted American independence. Lord Shelburne was generous to America so that France would have less of an influence there. General Washington would submit his resignation from the army after the war, and he instantly became a national hero. Washington and his men had truly achieved the impossible and won his country's independence. After a few celebratory stops in Baltimore and Annapolis, Maryland, Washington returned to Mt. Vernon and became a farmer once again.

As time passed, Washington made it known that he was an abolitionist, and he wanted an end to slavery (pg 213, #1). Though Washington did not take the steps of either Jefferson or Madison in the direction of religious freedom, he was equal in their assessment of creating a special, aristocratic government. Washington was a liberal conservative who had initially opposed the Articles of Confederation (pg 219, #1). The north and south of America seemed to be split on issues such as slavery. Washington did recognize the importance of the union of the states as a whole and opposed the fracturing of America into three factions: (1) the north, (2) the middle, and (3) the south.

According to Washington, to have an America that had a future of profit and stability, it was crucial to create "an indissolvable union of states under one federal head" (pg 221, #1). However, the Articles of Confederation were insufficient to create a strong national government. Washington himself aggressively undertook to arrange a convention in Philadelphia in May 1787, to "render the Constitution of the federal government adequate to the exigencies of the Union, and to report such an act for that purpose" (pg 223, #1).

Washington was appointed by the state of Virginia to meet in Philadelphia at the convention, but this was against his wishes, due to his suffering from rheumatism. Yet Washington's very presence added strength and credibility to the meeting. Washington was also a part of the original Declaration of Independence signing in Philadelphia in 1776, but Madison was given the title "father of the Constitution." The Constitution of the United States dictated that the federal government consist of three branches: the Executive, the House (including the Senate), and the Supreme Court. This gave the Constitution the concept of checks and balances so that no one branch could have too much power over the other.

At the time of its construction, however, it was lacking a Bill of Rights. It took Federalist leaders, such as George Washington, Benjamin Harrison, George Mason, and Richard Henry Lee, to decide to add a Bill of Rights section to the Constitution, in order to satisfy proponents on all sides of the issues (pg 232, #1). This act brought in the votes of the states that were unwilling to ratify the document without a Bill of Rights, such as New Hampshire, New York, and finally Rhode Island in 1790.

When the time came to choose a President, George Washington was the obvious choice among early American leaders. Surprisingly, Washington himself was reluctant to accept the position, having felt that he had already served his country, and enjoying his life as a farmer at Mt. Vernon. Despite this, Washington did accept the Presidency out of loyalty to his country, and he felt that he had to put duty first.

Washington was sworn in as President on April 30,1789, while John Adams was sworn in as Vice President, due to his coming in second in the Presidential race. In fact, it was Adams who proposed that Washington be addressed as "his Highness, the President of the United States, and protector of liberties." Washington saw the wisdom of declining such a title, and simply wanted to be called "President of the United States."

Washington chose an excellent Cabinet that included Thomas Jefferson as Secretary of State, John Jay as Chief Justice of the Supreme Court, and Alexander Hamilton as Secretary of the Treasury. Jefferson believed that individual rights were more important than national power, except in matters of foreign relations (pg 240, #1). Jefferson had gained experience as the American Minister to France before joining Washington's Cabinet.

John Adams, on the other hand, was somewhat diminished in his role as Vice President, which only allowed him to be the tiebreaker during a House vote. Adams was a champion debater who loved to passionately argue his point of view in front of his colleagues, and must have been frustrated in a role with such severe restrictions. Adams, however, put his country's needs first.

Washington's two key advisers in his Cabinet were Alexander Hamilton and Thomas Jefferson, who were polar opposites in their political beliefs. Washington used these conflicting points of view to his advantage by taking what he considered the best points of view from these two men and balancing them into his own ideology. For example, during Washington's first term as President, Article 12 was passed (Jay's Treaty), which opened U.S. ports to Britain, but not vice versa. Hamilton supported this idea as a way of giving England a concession after the war, so that the United States could still trade with England. This was in order to get America's economy up and running within the global market. However, Thomas Jefferson strongly disagreed with this idea, for he was in favor of giving special diplomatic privileges to France.

Nevertheless, Washington signed the treaty, for he cited that it was the Executive Branch of government, along with the U.S. Senate, that was in charge of foreign relations, as was guaranteed by the Constitution (pg 237, #4). Even though Jay's Treaty was passed into law, Washington continued to consider France more of an ally to America than England (pg 160, #4).

Other problems during Washington's Presidency, besides the trouble from the Native American attacks, included the national debt problem and the Whiskey Rebellion that took place in western Pennsylvania. Washington felt a tax was needed to create federal revenue. After some fighting and deaths that resulted from this clash, the whiskey tax was accepted and enforced. It could be contended that a bank would supply a sound currency and useful loans to the federal government, which it did.

Though Washington's first two years were considered successful, the President was still having problems with Britain, Spain, and the Native Americans, while migrating out west into the frontier. Washington skillfully negotiated with these entities to avoid conflict. During this time, France was going through a revolution, resulting in the deaths of King Louis the 16th and many French civilians. As the Presidential election of 1792 was looming, Washington decided that he would retire and not seek office for a second term. However, all of his associates begged him to stay, knowing the respect and admiration that his name brought. This was certainly needed in 1791, when Native Americans attacked American regiments in the old northwest region and inflicted hundreds of casualties to U.S. soldiers and civilians (pg 255, #1).

President Washington chose General Anthony Wayne to lead the charge of troops down the Ohio River towards Ft. Washington. After some battle and diplomatic negotiations, the Native Americans recognized their defeat and signed a treaty with Wayne at Greenville in 1795 that opened eastern and southern Ohio and a small part of Indiana to white settlement (pgs 258-59, #1). Eight years later, Ohio became a state. Originally, America favored the French Revolution, until the French murdered citizens of organized Christianity, which was especially offensive to New England Federalists.

With ever-mounting pressure on Washington to either go to war with Britain or France, Washington managed to avoid both by showcasing his brilliant fusion of advice and ideas from both John Adams, who supported England, and Thomas Jefferson, who supported France, and successfully maintaining a policy of strengths of both arguments. By avoiding war against Britain or France, Washington decided to finish his Presidency in 1796, with an elected successor to replace him. Washington's modesty with power served as an example for all future presidents by rejecting the notion to be a king or an emperor, which has greatly served the strength of this country and for which we can truly be grateful.

Washington suggested that the nation should avoid "overgrown military establishments, particularly hostile to Republican liberty" (pg 288, #1). Washington also said that America should have a strong central government, in order to have a greater security system against a foreign attack. According to Washington, the American government should cultivate peace and harmony with all nations. His vision, all those years ago, proved to be brilliant foresight that evolved into a true and fair democracy. As we have progressed through the following two centuries, we have seen the benefits of having two or more points of view expressed freely to summarize the entire population of the nation. Washington was ahead of his time by seeking to solidify the Union above all else, despite its differences in regional and cultural life (pg 288, #1).

Washington also had the wisdom to openly examine his mistakes, which he hoped his country "would never cease to view" (pg 289, #1). This was stated by Washington so that his beloved country would create good laws under a free government. In the end, Britain retreated to Canada and America had a peaceful relationship, for the time being, with the Spanish and Native Americans. Washington arranged to have all of his slaves freed upon the death of his wife, Martha. George Washington died on December 14th, 1799, as one of the greatest heroes that America would ever know. Washington seemed to conquer all adversity that was dealt to him in life, and his legacy only proves his resilience, incredible leadership abilities, and brilliance.

Bibliography

1. *George Washington: A Biography*, by John R. Alden (Louisiana State University Press, 1984), disc 2.
2. *Adams: An American Dynasty*, by Francis Russell (American Heritage Press, 1976), disc 2.
3. *Affairs of Honor*, by Joanne Freeman (Yale University Press, 2001), disc 4.
4. *Patriarch*, by Richard Norton Smith (Houghton-Mifflin, 1993), disc 4.
5. *George Washington's War: The Saga of the American Revolution*, by Robert Leckie (Harper-Collins, 1992), disc 5.
6. *The Wars of America*, by Robert Leckie (Castle Books, 1992), disc 6.
7. *Encyclopedia of the American Revolution*, by Mark M. Boatner (D. McKay Co., 1966, 1974).
8. *1776*, by David McCullough (Simon & Schuster, 2005), disc 10.
9. *The Complete Book of U.S. Presidents*, by William A. Degregario (Barricade Books, 2001).
10. *The American Revolution*, by Colin Bonwick (University Press of Virginia/Macmillian Education, 1991).
11. *The Revolution* (History Channel DVD).

INTERVIEW WITH CECELIA DESMOND

Thomas Athridge: Was Washington born into a life of privilege, or did his family earn their way to wealth?

Cecelia Desmond: Washington's father had farms and plantations, but he wasn't one of the privileged elite of Virginia. George Washington inherited Mt. Vernon, and once he married Martha, and got some cash, and through the use of her money, gradually turned it into the magnificent plantation that it became.

TA: How damaged was Washington's reputation by his overwhelming defeat at Ft. Necessity during the French-Indian war?

CD: Washington is credited with starting the Seven Years' War through his actions at Ft. Necessity. However, the overwhelming defeat he suffered during the French and Indian War wasn't at Ft. Necessity, and it wasn't Washington's defeat, it was General Braddock's. Washington was in total control of rallying and saving Braddock's army. Washington was utterly fearless during the battle—his cloak was shot full of holes and several horses were shot from under him, but he stayed at the front, keeping the troops together.

TA: How damaging would a relationship and possible betrothal have been to Sally Fairfax?

CD: Washington was very friendly with the Fairfaxes—they mentored him. A relationship beyond friendship with Sally would have been a betrayal of the Fairfaxes' trust in him, and a betrayal of Washington's own character.

TA: After marrying Mary Custis and becoming a farmer, how successful was he in producing crops and accumulating profits?

CD: He was a successful farmer. He was very successful in diversifying his crops rather than relying on tobacco, as many did, during colonial times. He grew grain for mills and started a fish business.

TA: Could the Hessians be considered terrorists?

CD: No, they were military troops hired by the British. The mark of a terrorist is use of violence and includes killing civilians as targets. Hessians were brutal , but went after military targets (although they committed a number of rapes, especially around Trenton, New Jersey). If anything, the Sons of Liberty were the terrorists, and not the Hessians. The Sons of Liberty tarred and feathered those they didn't like, and tore down [Thomas] Hutchinson's house with one of the best libraries in America.

TA: Would the war have been won without French assistance?

CD: Whether the Americans would have won without the French is impossible to know, but my guess is that the Americans would have eventually won, but that the United States would not include the South, which was in British hands.

TA: Was Washington's combat versus the Iroquois similar to our current war versus terrorism?

CD: The Americans conducted a war of terrorism on the Indians under Washington's direction. General Sullivan led the war on the Indians and was ordered to attack in late summer and early fall and destroy the Indians' crops to make it impossible for them to live through the winter. Daniel Morgan's riflement were Indian fighters before the war—they joined the army because they were told they could kill Indians legitimately.

TA: Did Washington want to abolish slavery?

CD: Washington arranged for the emancipation of his own slaves after his death.

TA: How wise was Washington's philosophy on a democratic government as opposed to a monarchical government, in which he would have been King?

CD: Washington was much beloved by the American people and by his army and could have taken over the civilian government as king if he so chose. He did not. He always deferred to the civilian government during the war and voluntarily withdrew from the army after the war, never taking advantage of his inherently powerful position. He set the stage and became the model for the rest of American history with respect to government officials being elected to their positions, and the civilian authority controlling the army.

TA: What other books on George Washington would you recommend?

CD: *Life of Washington*, by John Marshall; also, *A Biography of George Washington*, by George Washington Irving.

TA: Based on the code of honor between leaders of Washington's time, did dueling and other acts of honor give political credibility or humility? For example, the Hamilton-Burr duel.

CD: Generally, when challenged to a duel, one was honor bound to participate. Also, generally, the men would not aim to kill, and all could go home satisfied. In the Hamilton-Burr duel, Hamilton aimed away from Burr, but Burr apparently aimed to kill Hamilton, and of course he did. Burr's reputation suffered because of this.

TA: Based on your knowledge of President Washington, if he were alive today and in power, how do you think he would handle the present war versus terrorism, based on his history?

CD: If Washington were alive today, he would meet force with both force and diplomacy, and he would not have tolerated such an attack on Americans. He probably would not confer too much with other countries before retaliating, since he thought we should avoid "foreign entanglements."

INTERVIEW WITH JERRY GAWALT, LIBRARY OF CONGRESS EXPERT ON GEORGE WASHINGTON*

Thomas Athridge: Was Washington born into a life of privilege, or did his family earn their way to wealth?

Jerry Gawalt: His father was a wealthy plantation owner and came from a prestigious family.

TA: Was it unusual for a 20-year-old soldier to be appointed a major?

JG: Without Washington's political influence, he would have not a chance.

TA: How damaged was Washington's reputation by his overwhelming defeat at Ft. Necessity during the French-Indian war?

JG: His reputation was actually enhanced, but he also had a reputation of being lucky in battle and had an account of his campaign published.

TA: How damaging would a relationship and possible betrothal have been to Sally Fairfax?

JG: It would definitely have been a scandal, and papers on Washington's thoughts can be found in the Washington papers to Sally Fairfax on loc.gov.

TA: After marrying Mary Custis and becoming a farmer, how successful was he in producing crops and accumulating profits?

JG: He was a very successful farmer at Mt. Vernon.

TA: Did the Stamp and Townshend acts anger Washington enough to revolt?

JG: Washington was deeply troubled by the Stamp-Townshend acts because England gave no fair opportunities to advance. Washington was also upset over the treatment England showed during the French-Indian War that Washington participated in combat.

TA: How important was the first Continental Congress in picking who the leaders were and declaring independence?

JG: Washington was picked in this crowd of independence leaders because he was the best.

TA: Did Washington and John Adams see eye to eye or differ in governmental issues in regards to the revolution?

JG: They were equally radical. At the time, they were not political rivals, but Adams was also pro-British.

TA: After signing the Declaration of Independence, did Washington realize the importance of the framework of democracy that he helped create, or was he more concerned about secession from England?

JG: He did not sign the Declaration of Independence, and the country was not a democracy at the time, but a conservative government.

TA: Could the Hessians be considered terrorists?

JG: Yes, Hessians were terrorists.

TA: How high was the tension between Washington and Adams when Adams demanded that high officials be elected every year to show that the military answers to Congress?

JG: Washington wanted a standing army; Adams wanted a militia army.

TA: How important was Washington's diplomatic skills in convincing France to fight England?

JG: John Adams and Benjamin Franklin did most of the work.

TA: Would the war have been won without the French?

JG: Unknown, but not a deciding factor.

TA: Was Washington's combat versus the Iroquois similar to our war on terrorism?

JG: American treatment of the Indians was bad. [John] Sullivan burned the Iroquois, and the army went after the Indians.

TA: With the departure of French General D'Estaing to France, how did General Nathanael Greene attempt to save a South in mostly British control?

JG: The 1st and 2nd arrivals (Cornwallis, Greene directly involved) of partisans in the South was more terrorism.

TA: What did Washington learn from the shocking betrayal of Benedict Arnold?

JG: Arnold and Washington did not get along, and Washington did not trust him. Division of command is what Arnold wanted but did not get.

TA: How did Washington keep his men alive and not mutinying during the winter of 1777-78?

JG: Because Washington stayed there himself with his men, and shared the experience with them, along with Martha. He lobbied Congress for more money, and when he had to, he simply confiscated food from Pennsylvania farmers. This was because the British had more money to buy things.

TA: Did Washington want to abolish slavery?

JG: Yes, Washington did want slavery done. He was the only president to act on it, yet was somewhat hypocritical. Martha freed all of his slaves.

TA: Did Washington really want the presidency, or did he just feel obliged to serve?

JG: Yes, he very much did want to be president, and he was eager for it.

TA: Did Adams as vice president provide a counter-balance of political ideology?

JG: No, just a tie-breaking vote. The president and vice president never consulted [each other].

TA: How influential was Washington in designing a bill of rights?

JG: He had no part in the bill of rights, and did not see them as a necessity. The courts decided the rights for the people.

TA: How diverse was Washington's Cabinet in ideas?

JG: He liked geographic diversity and wanted his Cabinet to represent people throughout the country, but it was not his idea to have Jefferson and Hamilton disagree so much.

TA: How wise was Washington's philosophy on a democratic government as opposed to a monarchical government, in which he would have been king?

JG: Washington was in favor of a republican government, and that educated people had a say and no one else—no women or black people or Native Americans. The presidency is not that different now than then, and he was opposed to a dictatorship. Democracy differs from republicanism.

TA: What other books would you recommend?

JG: Abraham Flexner books, Charles Royster. See the Washington papers at the University of Virginia, and Philander Chase.

TA: Based on the code of honor between leaders of Washington's time, did dueling and other acts of honor give political credibility or humility?

JG: Yes, Washington did believe in a code of honor, and the gentlemen's code, though he was never in a duel.

TA: Based on your knowledge of President Washington, if he were alive today and in power, how do you think he would handle the present war versus terrorism, based on his history?

JG: Washington would attack the enemy much more than Bush is doing. Washington was an extremely aggressive leader and sentenced colonial troops to hanging, and sent his army to destroy Indian villages.

*Gawalt recommends looking at Thomas Jefferson's handling of the attack of the Barbary pirates and Anthony Pitch's *The Burning of Washington.*

Chapter 3
The Barbary Pirates War and President Thomas Jefferson's Actions During This Conflict

As history would record, the first president to launch a war on foreign soil would be our third president, Thomas Jefferson. History would normally regard Jefferson as a passive leader whose vision of equality was fundamental to American democracy. Yet it was Jefferson who chose to fight the Barbary pirates, who were a group of terrorists from the Barbary States who attacked ships at sea, and inflicted their reign of terror on innocent civilians and commercial shipping. These raids were being conducted by the four Barbary states: Tunis, Algeria, Morocco, and Tripoli (now Libya).

The Barbary pirates were an early form of Islamic terrorists, in that they would capture vessels at sea, in the name of Islam, then extort as much money as possible from the captured vessels' governments and send the prisoners into repressive prison systems. During this period, from 1500 to 1800, piracy was a common practice in the Mediterranean Sea, even going as far back in time to Julius Caesar himself, who was attacked by pirates (pg 29, #1). These 4 Muslim-controlled entities had openly declared an Islamic jihad, or holy war, versus Christianity and all infidels.

The word "jihad" is from the word "jihada," which means "to strive." (pg 8, #3). Yet these jihadists were misinterpreting the information from the *Koran*, which clearly states that jihad is non-violent and only to be used as a means of self-defense. This clear abuse of the term jihad was regularly practiced by the bashaws, deys, and beys, who controlled the Barbary states then. But in order to understand how this piracy situation became so uncontrollable, and how the United States became directly involved in declaring war on these savages, first we must go back to Spain in the year 1492.

In 1492, King Ferdinand and Queen Isabella of Spain had sent Christopher Columbus to the new world, which turned out to be America instead of India. The Spanish government had the intent of spreading Christianity throughout the new world. This act of spreading Christianity would bring them in direct conflict with the Moors, who were the descendants of the Islamic conquerors of Iberia in 711 A.D., who also lived in Spain (pg 9, #3). As a result, the Christians and Muslims fought a holy war in Spain. The Christians beat the Moors, and King Ferdinand ordered the defeated Moors to either be baptized or expelled from Spain (pg 12, #3). When the defeated Moors refused this order, the Christians persecuted the Muslims.

This persecution forced the Muslims into exile from Spain, and these 3 million refugees would eventually settle in the 4 countries of Tunis, Algiers, Morocco, and Tripoli (pg 14, #3). These states would become the Barbary kingdoms, named after the Greek word for "barbarian." These four states unleashed a horrifying and destructive path of piracy and terror that lasted for over 300 years.

As a reward for capturing and persecuting the Christians, the Muslims believed that they would go straight to paradise, called "houris." (pg 30, #1). These pirates were successful in capturing hundreds of ships and enslaving the prisoners of all nations until their ransom had been paid (pg 18, #3). But the nation that the Pirates were the hardest on was Spain. In fact, by 1634, the nation of Algiers had 25,000 Christian slaves in its possession, who were mostly Spanish, Portuguese, and Italian (pg 20, #3). These slaves were mostly white.

The pirates were quite cruel to the Christian slaves. For example, the prisoners would be subjected to such extreme acts of brutality as being roasted alive or crucified. The Muslim captors perpetrated these extreme punishments on the prisoners for blaspheming the *Koran* (pg 31, #1). The prisoners were treated like animals and were often killed, though they would be worth more money to the pirates alive than dead. In fact, the white slave trade would be very prolific and profitable for the pirates from 1575 to 1769 (pg 25, #3).

Europe was not able to properly fight the pirates at the time, because they were too busy fighting each other. For example, Holland had finally won its independence from Spain in 1648 after 80 years of war, and 1648 was also the end of the Thirty Years' war, which began in 1618. By 1591, the Ottoman-Barbary association had split. Only when the Barbary piracy would get out of hand would countries such as Holland, France, Spain, or England occasionally send military forces to Algiers and Tunis and unload massive bombardments upon these pirate states. These military strikes would end up killing thousands of pirates, but never to the extent of fully wiping them out (pg 29, #3).

Holland, which had finally grown sick of dealing with the pirates, had agreed to pay an annual tribute to the pirates to stop the attacks. The government of Holland sent the pirates $5,000, along with Dutch cannons and swords, which, ironically, were the very weapons that the pirates would later use against them. Following Holland, Austria, Sweden, and Denmark paid the pirates an annual tribute to cease the attacks. Yet the pirates would break their promise and continue the attacks at sea (pg 30, #3).

By this time, America was fighting England for its independence, which was officially won in 1783. Because of its unique status as a new nation, the United States had signed no treaty for ransoms with the pirates, which left them susceptible to attack. In response to this problem, U.S. mediator John Adams met with the Tripolitan ambassador Abdrahaman, who told him that unless the new U.S. government paid the forced tribute placed upon them, Tripoli would be at war with America. The price of peace, according to Abdrahaman, would be $66,000 a year. Yet this payment would only be for the country of Tripoli The other 3 countries (Tunis, Morocco, and Algeria) would demand their own prices, which would each be different and more difficult, according to Adams and Jefferson. .Abdrahaman's argument for these ransoms was that pirating rights were guaranteed by the *Koran*, which is untrue. This excuse was also offered by the other Barbary states for the monetary demands they were seeking in order to stop their own shipping attacks. When Thomas Jefferson became president in 1801 and added the total numbers of dollars the 4 individual Barbary states were seeking, the amount was a staggering $1.3 million.

Jefferson thought that the United States should not pay this amount to the pirates, yet John Adams thought that the U.S. government should, in order to help U.S. commercial shipping remain free of Barbary attacks, which was important to the economic structure of the new republic. Adams did believe, however, in the U.S. government's strong commitment to constructing a new navy. Based on the evidence, Jefferson had established 5 reasons as to why the U.S. armed forces should fight the pirates:

1. Justice favored this opinion
2. Honor also favored this opinion
3. It would procure the United States respect in Europe, and respect is a safeguard for one's interests
4. It would arm the federal head with the safest of all the instruments of coercion over its delinquent members, and prevent it from using what would be less safe measures
5. It would ultimately be the least expensive option (pg 44, #3).

This was especially true because American trading was ripe for European consumers, with U.S. vessels eager to trade goods, such as flour, sugar, fish, fruits, olive oil, and opium, and to establish capital (pg 33, #1). In 1785, 2 U.S. ships were captured by the Algerians, the *Maria* and the *Dauphin*. The hostages of these ships were tortured and auctioned off. The prisoners were treated like animals by their captors. The captain of the *Dauphin*, Richard O'Brien, wrote back to America, pleading for help. Because of the limited resources, there was little the Continental Congress could do, due to the lack of federal funds to meet the Barbary ransom demands. By the time George Washington took control of the presidency, Algiers was demanding $59,496 from the

United States for both captured crews of the *Dauphin* and *Maria*. These attacks would mercilessly continue.

On October 25th, 1793, the U.S. ship *Polly* was captured by the Algerian pirates, led by Rais Hadga Mahomet. This act was followed by 10 additional seizures of U.S. ships by Algiers, which would total 119 captured American prisoners. The prisoners were treated like dogs and forced to work in labor camps. Needless to say, the prisoners were treated inhumanely by the evil pirates (pg 64, #3).

One of the most brutal Barbary slave leaders was a man named Sherief. During his brutal reign as prison abuser, he would not allow the prisoners to be fed during the holy month of Ramadan. Because of this, Washington, as president, saw the dire need of the U.S. armed forces to build a navy. Consequently, by the end of 1793, the U.S. Congress debated a plan to build six ships, at the cost of $688,888. President Washington officially signed this into law on March 27, 1794. The Secretary of War, Henry Knox, was responsible for building the navy from scratch, with assistance from American sea captains, businessmen, and ship builders (pg 70, #3).

Accordingly, March 27th, 1794 is considered the launch of what is now the U.S. Navy and Marine Corps, though both services existed for nearly twenty years prior (pg 35, #1). The 6 new ships that were constructed were the *United States, Constitution, President, Constellation, Chesapeake*, and *Congress*. In fact, the ship *Constitution* would be nicknamed "Old Ironsides."

The Algerian leader at the time, Dey Ali Hassan, was a serious adversary, and was widely considered the strongest of the Barbary pirates. Dave Humphreys and Joel Barlow were the negotiators with Hassan for the United States. The dey demanded $2,247,000 and 2 ships worth $248,000 to be given to him in order to have a treaty in 1795 (pg 75, #3). After much discussion and negotiations, it was eventually agreed that the United States would pay Hassan $642,000, plus an annual tribute of $21,600. The treaty was signed on September 5th, 1795, which cemented payments by the United States to Algeria, Tunis, and Tripoli (pg 37, #3). Unfortunately, the pirates continued the attacks on shipping, because the pirates wanted to escalate the prices for peace as time progressed. . Washington approved of the treaty stipulations in 1795. However, Jefferson certainly did not. In fact, Jefferson did not want to pay the ransom at all, as the pirates continuously broke the signed agreements and regularly increased the prices they demanded.

The situation got so bad that the ship *George Washington*, whose captain was William Bainbridge, was forced to transport Mustafa Bobba's presents and bribe money to the Ottoman capital of Constantinople. If he declined, he would be attacked by the dey's forces. When word of this atrocity reached the United States, both Madison and Jefferson were outraged, and they knew that something had to be done (pg 99, #3).

In the meantime, James Cathcart, who was the U.S. envoy to Tripoli, was receiving hostile demands from the Tripolitan leader Yusef. By January 3rd, 1801, Tripoli sternly warned the United States that until they were paid, there would be imminent war between the 2 countries (pg 100, #3). The price of peace that the bashaw of Tripoli set was $225,000, plus an annual tribute of $25,000. The bashaw demanded this because other countries such as Sweden, Holland, and Denmark were already paying that price for peace.

Yusef was expecting a sizeable treaty from the United States under President Adams in 1800-01, but the situation changed when Jefferson assumed the presidency in March 1801 (pg 104, #3). When President Jefferson examined the Barbary threat situation, he deemed that war with the pirates would be the preferable option, because the U.S. federal government could not afford the continued escalation of demands from the Barbary states. In order to understand why President Jefferson chose this option of war against the Barbary countries, first we must closely examine the character of Jefferson the man and the leader, what led him to his position of great influence in the founding of the United States of America, and how Jefferson came into the position of being the first president to successfully land an American military force on foreign soil.

Thomas Jefferson was born on April 13th, 1743, at a location called the Shadwell plantation in Albemarle County (formerly called Goochland County), Virginia. His parents were Colonel Peter and Jane Randolph Jefferson. His father Peter was a surveyor, planter, public official, sheriff, justice of the peace, and court judge. As a student, Jefferson thrived in Greek, Latin, and French. In 1760, Jefferson enrolled at the College of William and Mary in Williamsburg, Virginia. He excelled in science, philosophy ,mathematics, and literature while a student there. After 2 years of college, he enrolled to study law under George Wythe, and he was admitted to the Virginia Bar in April 1767.

He believed in deism, under the Anglican banner, yet he believed in a creator that was mostly uninvolved with the affairs on the earth (pg 40, #4). He did believe in the doctrines of Jesus. Above all his other political opinions, Jefferson firmly believed in the separation of church and state. Jefferson was not questioning whether God existed, but wanted the complete freedom of the citizens of the United States to worship as they saw fit. In other words, all religions in America would be welcome.

Jefferson married Martha Wayles Skelton on January 1st, 1772, in Charles County, Virginia. He became a member of the House of Burgesses from 1769 to 1774, until his heavy involvement in the First Continental Congress in 1774. Later, in 1776, Jefferson was a principal author of the Declaration of Independence, with input from Ben Franklin, Robert Livingston, John Adams, James Monroe, James Madison, and Roger Sherman. Yet this was clearly a Jefferson product, based on ideals of a free and equal society, with religious freedom and education. During the

Revolutionary War, Jefferson was a member of the Virginia House of Delegates from 1776 to 1779. In 1779, he was elected Governor of Virginia during some difficult years in the enormous struggle of the Revolutionary War. During Jefferson's term of 1779-81, he had to abandon the state capital of Richmond due to a British raid on the city, and was also forced to flee his home of Monticello by invading English troops. Jefferson's years as governor could be largely seen as unsuccessful and trying. Yet Jefferson's leadership skills were improving, and his influence was becoming enormous in the inner circle of American political power.

Jefferson loathed oppression, which became the main theme behind his political ideology (pg 50, #4). As brilliant as Jefferson's writing was, he did have difficulty speaking in public. Yet his friend John Adams could speak very well in public. The two would later become political rivals, but retained a strong friendship in their post-presidential years.

Tragedy befell Jefferson in 1782.After almost being captured by the British in Virginia at Monticello, Jefferson's wife, Martha, died on September 6th, 1782, after becoming fatally ill following the delivery of their child, Lucy Elizabeth. Lucy Elizabeth herself would also later die from a bout with whooping cough. These tragic events in Jefferson's personal life left him terribly shaken.

After these devastating losses in Jefferson's life, he developed a personal bond with James Madison and James Monroe. These men shared Jefferson's vision of personal and political freedom, and these 3 men occupied the White House for the first 24 years of the 19th century. Yet he also shared a close bond with his future political opponent John Adams while the two men were in France, where Jefferson was appointed Minister in 1785. Both Adams and Jefferson disliked George the III (pg 88, #3). Jefferson did like the French people, and he favored them in disputes with England. Madison agreed with Jefferson.

The combination of Jefferson's and Madison's ideas would form the very structure of our American government, with Jefferson's idea that "the will of the majority should always prevail" and Madison's passionate beliefs in social balance (pg 122,-#3). Jefferson also believed in limitations of the Executive Branch, based on what he saw as the clear abuse that monarchy rule had created in Europe.

When Jefferson became President of the United States in March 1801, the Barbary states continued threatening U.S. commerce and passenger safety and represented a menace that the president had to directly deal with. So that is precisely what Jefferson did. He had decided on his own authority, which the president has the right to do, to send the ships *Essex*, *President*, and *Enterprise* to the Mediterranean to directly monitor the pirates. Jefferson needed no declaration of war to do this.

Jefferson, once he was president, formulated three options to deal with the situation:

1.   Pay the annual tribute

2. Deal with the ransoms when people were captured
3. Use force to deal with adversaries, once and for all
Yet the problem was that the Naval Reduction Act reduced the defense budget by $500,000. This act also reduced the number of ships from six to four. These four ships possessed 124 guns combined. At the time, the U.S. Marines had approximately 350 men (pg 111, #3).

The Barbary pirate named Murad Reis was once a Scotsman named Peter Lisle, who was captured on the vessel *Betsey* in 1796 (pg 112, #3). While he was in captivity, he converted to Islam. Tripoli then had a Muslim population of approximately 30,000. Jefferson did not feel that it was right for the United States to pay the annual tribute to the pirates, because countries like England, France, Denmark, Sweden, and Venice were all paying the pirates a lot of money to begin with. On March 9th, 1801, five days after his inauguration, Jefferson met with his Cabinet to form a military plan to fight the pirates. Jefferson wanted the U.S. government to build faster, stronger ships with more firepower to legitimately fight the pirates. The U.S. Naval squadron had blocked Tripoli from the sea and was ready for a fight.

On August 1st, 1801, the first naval battle of the Barbary War occurred when the ship *Enterprise* took on Murad Reis's ship, the *Tripoli*. The *Enterprise* won this battle convincingly by killing 30 of the 80 officers on board the *Tripoli*, while sustaining zero casualties. Because of this, Admiral Reis returned home in total disgrace, and his superior, Bashaw Yusef, stripped him of his command and had Reis beaten with a bastinado (whip) and tied to a donkey.

Jefferson was impressed by the victory of the *Enterprise*, but in order to officially declare war on these Barbary states, Jefferson knew that it had to be done constitutionally, by the book. This meant going to Congress to officially declare war. Because the bashaws were receiving gifts from all of the other countries in Europe, they could turn their full attention to America as their only enemy. To counter this, on February 6th, 1802, legislation was passed that protected regular ships with the U.S. Navy. Also, President Jefferson ordered all ships that were seized from Tripoli to be made prizes for the United States (pg 56, #1).

In 1803, Congress gave the navy $50,000 to build up to 15 gunboats, and an additional $96,000 for four warships (pg 136, #3). Jefferson named Edward Preble to command U.S. forces against the pirates. Preble had served the American forces in the Revolutionary War, and became a P.O.W. by being captured by the British ship *Jersey*. While a prisoner of the British , Preble was struck by typhoid fever (pg 62, #1). Preble had shown sound military command to Jefferson, and the President believed that Preble was the answer for the United States to stop paying extortion money, once and for all.

One bad episode in this Barbary War occurred when U.S. commander William Bainbridge got his ship, the *Philadelphia*, stuck on a reef off the shore of Tripoli, which resulted in the capture of the ship and its crew (pg 162, #3). Bainbridge's mistake was his decision not to wait out the low tide and escape during the high tide. When the crew was captured, Preble was successful in forwarding thousands of dollars through the captured negotiator, Cathcart (pg 180, #3). Of the men who were prisoners of the pirates, the officers were treated much better than the enlisted men, who were treated as badly as animals. With the knowledge of this prisoner abuse taking place, American leaders Stephan Decatur, William Bainbridge, and Edward Preble suggested the destruction of the captured ship in Tripoli Harbor (pg 181, #3).

On December 23rd, 1803, a secret mission to rescue hostages and reclaim the ship *Philadelphia* from the pirates took place at night, led by Decatur, Salvatore Catalino, and Preble. They made their attack on board the ship *Intrepid*. The *Intrepid* was once the pirate ship *Mastico*, which was successfully captured by the Americans in a raid. Stephan Decatur commanded the *Intrepid*, while Lieutenant Charles Steward commanded the 16-gun *Siren*. While these ships made their attack on the *Philadelphia*, different officers were assigned to different sections. When the attack took place, the U.S. troops either killed or made everyone who was onboard the ship flee. The *Philadelphia* was successfully back in American hands. Next, Decatur ordered that the men set fire to the *Philadelphia* while they could escape to the *Intrepid*. When the fire was set, the enemy fired shots from the shore, and unsuccessfully at the *Intrepid*. In a miraculous turn of events for America, the cannons from the burning ship *Philadelphia* discharged rounds at the pirates on the shore before finally exploding at sea. Their mission was a complete success, and the evacuating Americans had completed their mission in an astounding 25 minutes. This act hurt the Barbary cause significantly (pg 194, #3).

Though both the Pope and the countries in Europe were pleased by the American victories, U.S. prisoners in Tripoli were punished even harder as a result. Despite this, President Jefferson promoted Decatur to port captain for his victorious results. Decatur was the youngest captain ever for his time. Meanwhile, Commander Preble was anxious to attack Tripoli again, in an effort to wipe them out completely. He sailed to Naples, which was the Kingdom of Two Sicilies, to ask its leader for the use of some Italian gunboats. Preble was successful in obtaining eight gunboats and two bomb ketches from the Italians. Preble used these newly acquired weapons, combined with his own arsenal and battle plans, to mount an attack against Tripoli. Preble commenced this attack on August 3rd, 1804, with Murad Reis as Preble's Barbary opponent. Reis commanded two Barbary divisions. In this 10-minute battle, the Tripoli members suffered 52 killed and wounded men, compared to only six U.S. casualties. As a result, the Americans were the victors in this battle.

Despite this victory, Stephan Decatur's brother James was killed in this particular battle. When Decatur learned of his brother's death, he went berserk and launched a full assault on the Tripolitans onboard another ship. Decatur successfully killed the assassin and subdued the opposing army's crew.

Preble was known by his soldiers as a hard, strict disciplinarian, but the Marines under his command would greatly benefit from this. Preble would continue his assault on the Tripolitans on August 24th, 1804, by bombing the city of Tripoli. Preble wanted to inflict fear in the enemy's hearts, just as the Tripolitans had done to the Americans and other countries previously (pg 94, #3). As a result, the U.S. forces did some extensive damage to the Tripoli castles where the enemy was firing on the Americans, including Yusef's personal castle.

Despite these U.S. victories, the United States also suffered losses when a U.S. gunboat blew up in battle on August 7th, 1804, killing 10 and mutilating six. Also, the ship *Intrepid*, which did so well in the capture and destruction of the *Philadelphia*, was set to explode in a Tripolitan harbor, with the *Intrepid* crew safely evacuated and transported to another ship. Tragically, the *Intrepid* blew up earlier than planned, killing the crew and not doing much damage to the Tripoli harbor.

By this time, Preble had already been replaced by Commodore Samuel Barron. Preble was rewarded for his gallantry upon his return to Washington and was regarded as a hero for fighting bravely against a powerful enemy for the sake of freedom. While the United States continued to pursue the surrender of Tripoli, Commander William Eaton was in charge of putting a pro-U.S. leader named Hamet Karamanli back on Tripoli's throne. Eaton could speak four Arabic dialects as well as four Native American languages (pg 239, #3). Hamet was the brother of Tripoli leader Yusef, a sworn enemy of the United States. Though Hamet was ambiguous about aligning himself with the U.S. cause of pirate extinction, it was Eaton who was able to convince Hamet to join the U.S. cause to end all hostilities and defeat the pirates, in return for his being reinstated to the Tripolitan throne (pg 251, #3).

They agreed that the United States should head for battle at the city of Derna, which was Tripoli's second largest city (pg 280, #3). Derna's governor, and America's adversary, was a man named Mustifa Bey (pg 282, #3). Before launching the attack on the city of Derna, Eaton offered to Bey a position in Hamet's government if he chose to peacefully surrender. Bey's response to Eaton was, "My head or yours" (pg 282, #3). Even back then, the Muslim cry before going into battle was "Allahu Akbar," which means "God is great." Because of these unsuccessful talks, the United States stormed the city of Derna and won the battle. This was the first decisive land victory for America in the Barbary War (pg 284, #3).

A soldier named Captain Presley O'Bannon was given the title "Hero of Derne Tripoli, Northern Africa" for his heroic actions in battle against the pirates on April 27th, 1805, and he was credited as the very first U.S. Marine to plant the American flag on foreign soil. Eaton's troops were resupplied at a crucial stage of the Battle of Derna by the U.S. ship *Argus*. Together with the U.S. ships *Nautilus* and *Hornet*, this gave Eaton and his troops the firepower to achieve the Derna victory.

Meanwhile, Mustifa Bey was hiding in a mosque, and he successfully eluded capture. He fled Derna to join Yusef's troops. Eaton's troops were very loyal to him because of his heroic actions in battle. Yusef, sensing defeat in the war with the United States, surrendered on June 3rd, 1805, and he released the prisoners of the *Philadelphia* on June 10th (pg 299, #3). Yusef's own citizens turned on him and vowed to fight him. Yusef would keep Hamet's family in his custody until August 1807, and he would remain in power. In fact, Jefferson, after defeating the pirates, did not call for regime change in any of the four Barbary states.

Hamet's role was critical in the victory over the Barbary pirates, yet he would only receive $6,800 from the United States, and thus he seemed abandoned by the U.S. military. The United States did agree to pay Tripoli $60,000 to officially end the war, which it did. In total costs, the Tripolitan War was a $3.6 million affair for America.

Tripoli was renamed Libya in 1911 (pg 363, #3). During President Jefferson's second term, on June 22nd, 1807, the *Leopard-Chesapeake* incident took place when the British ship *Leopard* fired on the U.S. ship *Chesapeake*. This happened because the *Leopard* believed the *Chesapeake* was carrying 3 missing U.K. sailors. As a result of this attack, 3 U.S. sailors were killed and 18 were wounded. Jefferson responded by signing the 1807 Embargo Act, which barred all foreign countries from U.S. trade, including England. Jefferson chose this option instead of declaring war against England. Unfortunately for the United States, it lost $50 million in trade and hurt the country more than it helped. As a result of foreign trade being shut down, the three U.S. ships stationed in the Mediterranean were brought home.

Captain Preble died on August 25th, 1807. He was a tremendous influence on the American success in the Barbary campaign. The Algerian Treaty would not come until December 6th, 1815 (pg 359, #3). In this treaty, the United States received favored-nation status, with no tributes paid, and secured the release of all U.S. captives (pg 359, #3). The results of this successful military campaign were as follows:

1.   The United States fought for and obtained sovereign trading rights with foreign markets, and freed the world and America from piracy's tyranny.

2.   This helped strengthen the U.S. economy at home and abroad.

3.   This war gave birth and strength to the U.S. Marine Corps and Navy, so that America no longer had to depend on foreign navies for

military success, such as the French naval help it received during the Revolutionary War.

4. This conflict helped establish the U.S. fighting force that would later be tested and victorious against England during the War of 1812.

Jefferson strongly believed in an America that was free from tyranny, and he showed tremendous presidential strength in defeating his Barbary foes. This war caused the extinction of the Barbary pirates.

Bibliography

1. *From Sea to Shining Sea*, by Robert Leckie (Harper Collins/Random House, 1993), disc 4.
2. *American Sphinx: The Character of Thomas Jefferson*, by Joesph J. Ellis (winner of the National Book Award) (Random House, 1996), disc 5.
3. *Jefferson's War*, by Joesph Wheelan (Carroll and Graf, 2003), disc 7.
4. "Thomas Jefferson: The Pirate War: To the Shores of Tripoli," by Christopher Hitchens, *Time*, July 5, 2004.

Chapter 4
The War of 1812, and How President James Madison and the Nation Handled the Conflict

After the United States had successfully defeated the Barbary pirates, President Jefferson continued his role as Chief Executive until 1808, when his Secretary of State, James Madison, was elected as America's 4th president. Madison was elected to power during a turbulent and frightening period in American history. France, led by Napoleon, and England had declared war on each other. Each country was determined to eliminate the other country's influence on Europe and in the world.

As much as Napoleon wanted to invade England after he had broken the Treaty of Amiens, he realized that the British navy was just too powerful to easily conquer (pg 124, #4). The two countries of England and France had been at each other's throats going back as far as the French-Indian war, also called the Seven Years' War, from 1756-63. During the Revolutionary War, it was France that made the difference in America's victory over a very powerful England. American leaders, such as Washington, John Adams, Jefferson, and Ben Franklin, were instrumental in gaining France's support, which enabled American military victory.

However, Adams and Jefferson would soon disagree on the alliances between the United States and both England and France. Adams had signed a peace treaty on November 30th, 1782, with the British Empire that was in direct violation of orders from Congress and the Franco-American alliance (pg 265, #6). Though the French were angry with Adams, Ben Franklin interceded to act as a negotiator between the United States and France. Adams was seen by the French as controversial because of his habit of favoring England; however, the United States received a lot more property from England as a result of the Treaty of Paris than was thought possible (pgs 284-85, #6).

While Jefferson and Adams developed a bond while spending time together in France during the 1780s, Jefferson really liked France (pg 99, #2). Ben Franklin also struck up a fantastic relationship with the French. In fact, Jefferson's great relationship with France produced his belief that "the will of the majority should always prevail" (pg 122, #6). Madison agreed with Jefferson and was influenced by Jefferson's concepts of liberty and freedom. Madison would disagree, however, with Jefferson's mantra of a limited federal government, followed by limited executive power. Madison did concur with Jefferson's hatred for England. While Madison was Jefferson's Secretary of State, he was involved heavily with Jefferson's passage of the Embargo Act of 1807, which forbade the United States from trading with any other nation. This act was passed with the express intention of hurting Great Britain (pg 231, #7).

This act was passed in response to the sinking of the U.S. ship *Chesapeake* off the coast of Norfolk, Virginia, for refusing to be boarded. Four people were killed as a result, with 10 additional wounded (pg 127, #4). The French also presented a problem to America by illegally seizing U.S. ships at sea, which was occurring often. The United States seemed to be in dispute with potentially two different enemies, England and France, on a global stage.

Mindful of the problem with England and seeking to avoid hostilities, Madison repealed the Embargo Act on March 1st, 1809. However, the British responded to this by inciting Native Americans to attack white settlers. This was a vain attempt by Britain to stop the U.S. expansion westward; Britain promised to give the Native Americans their own land in exchange for the attacks (pg 139, #4). The British were to pay a bounty for American scalps from the Native Americans.

The territory the British had promised the natives was in the Great Lakes region. In order to lead the natives to fight the Americans, the British recruited the Native American warrior Tecumseh, a notorious white hater. Tecumseh was a Shawnee who had fought against the Cherokee, Creek, and Choctaw tribes. His hatred of whites stemmed from the murder of Tecumseh's father by white hunters when he was a boy. Tecumseh had created the Alliance of Five Nations after his wars with the Cherokee, Creek, Chicksaw, Choctaw, and Seminole (pg 11, #2). Tecumseh's intent was to declare war on whites and reclaim his homeland. He went with his forces to prepare for the assault on Americans in Detroit.

When Madison became president in March 1809, he proclaimed that the Embargo Act would be lifted against either England or France if they stopped molesting and interfering with U.S. commerce (pg 410, #6). Madison's Secretary of State, Robert Smith, and the government of England had then made peace with each other, so that trade between the two countries could continue.

France, in the meantime, continued to seize U.S. ships because Napoleon badly needed money in 1809, and these seized U.S. ships and cargo would provide millions of dollars for the French treasury (pg 432, #6). But France did agree to America's rules on Macon's Bill in 1810 that England refused, thereby giving France an advantage with Madison. Also hurting England was the repealing of the Erskine agreement between the United States and England in regards to the British attacking American

51

ships again.

Another problem for the Madison administration was the situation in Florida. Madison wanted Florida occupied by the United States, and not by Spain. Spain had the power to conduct trade with England while in Florida, while the United States could not. Also, the British were supplying the Native Americans with weapons in the region and encouraging attacks against white settlers located there. If a foreign power were to take possession of Florida after the Louisiana Purchase, this foreign occupation could be a direct threat to the United States. Something had to be done to cease the open aggression that was being inflicted on the American people and government. President Madison, who was deemed by some colleagues as the "Father of the Constitution," knew that in order to declare war against England, he needed approval from Congress.

During this inevitable build-up to war, Madison made James Monroe his new Secretary of State, and proposed to Congress to increase the size of both the army and navy. This is an interesting paradox, historically, because his mentor, Thomas Jefferson, believed in a limited army and navy. With tension mounting between the United States and England, Madison met with Congress to ask for a formal declaration of war on five counts:

1. The impressment of American seamen
2. The violation of American neutral rights
3. The naval blockade of the United States
4. The British refusal to revoke the orders of the Council of 1807, which barred all neutral trade with France and her colonies, and
5. The British incitement of Native American hostilities in the west (pg 12, #3)

Though Madison was thoroughly angered by England's aggression against the United States, he was also feeling the pressure from the war hawks in Congress, like Henry Clay, who were pushing for war against England. This was a way for Madison to strengthen his chances for a second term as president, because the war hawks would have found another candidate had Madison refused war against England (pg 177, #4).

Madison's war proposal to Congress occurred on June 1st, 1812 (pg 96, #5). The declaration was officially passed on June 18th, 1812. Britain, though already at war with France, was still considered the most powerful country on earth at the time (pg 181, #4). Despite this, President Madison knew that war with England was the only way to have freedom of the seas (pg 482, #6) This war would become known as "Mr. Madison's War."

The United States began the war with a weak and unprepared army. Also, the federal press began to severely attack Madison for engaging the U.S. military in this conflict. However, President Madison told England's Ambassador, Augustus Foster, "If the orders of the Council are repealed, and no illegal blockades are substituted for them, and orders are given to discontinue the impressment of our seamen from our vessels, and to restore those already impressed, there is no reason why hostilities should not immediately cease" (pg 504, #5). The five brigadier generals who began the war by leading U.S. forces were James Wilkerson (senior brigadier general), Wade Hampton of South Carolina, Joseph Bloomfield of New Jersey, James Winchester of Tennessee, and William Hull of Massachusetts (pg 2, #1). None of these commanders had ever faced the enemy before.

Detroit was where the war began because the area was considered vulnerable. British and Native American forces had always threatened any settlers in the region, and U.S. forces wanted control of Canada to further quell any British/Native American attacks. Also, the United States rushed into war in Canada without proper planning or supplies to successfully combat the enemy. America wanted control over the Great Lakes region. General William Hull commanded U.S. soldiers in battle at Detroit.

Madison's orders to General Hull were to move down Lake Champlain against the city of Montreal, while supporting U.S. thrusts into Canada from Detroit, Niagara, and Sackett's Harbor. However, Britain's counter-commander, General Isaac Brock, had thousands of troops stationed from Montreal to Detroit, awaiting the U.S. ambush. Also hurting the U.S. forces were the decisions of both Massachusetts and Connecticut troops not to participate in the battle (pg 184, #4). At the time, the U.S. military was in total disarray (pg 11, #1).

Madison wanted to conquer Canada before England because he thought that Canada's army was too weak to protect themselves, and the desire of England to exploit Canada was too great to ignore (pg 99, #5). As a result, Britain controlled all 3 lakes, and the United States was dealt a significant blow in the war. In battle, however, the United States did manage to kill General Brock. As a result of this loss, Madison sent in General William Henry Harrison to reclaim the area from British control, the same area that General Hull initially had control over, including the capture of 2,500 American men.

The U.S. did claim some victories during the year of 1812. For example, the U.S. ship *Constitution*, which was commanded by William Hull's brother Isaac, defeated the British ship *Guerierre*. Also, Stephen Decatur, well known from his battles against the Barbary pirates, won a victory from his ship, the *United States*, over the British ship *Macedonia* (pg 45, #3). On October 9th, 1812, U.S. Lieutenant Jesse D. Elliott and his crew of 124 sailors seized the British ships *Detroit* (formerly the U.S. ship *Adams*) and *Caledonia* (pg 49, #3).

Even though New England had been against the war with England, President Madison compromised his beliefs to please New England's interests by enlarging the U.S. Navy (pg 527, #6.). The Federalists believed in a strong navy. Robert Barclay, an American military leader, would use the dying words of his comrade James Lawrence when he was killed aboard the *Chesapeake* as a rallying cry: "Don't give up the ship!" (Note that this is similar to hijacked passenger Todd Beamer's last words, "Let's roll," onboard his hijacked plane on 9/11.)

During this time, there was a presidential election going on, and Madison successfully won a second term against his in-party opponent, DeWitt Clinton (pg 526, #6). Meanwhile, things were not good for U.S. prisoners in Detroit. General Henry Proctor had taken over command from the deceased General Brock and forced his prisoners on a march from Detroit to Montreal (pg 47, #3). Back home, former President Jefferson called Hull's surrender in Detroit "detestable." But Hull's replacement, future President William Henry Harrison, did much better in battle against England in Detroit.

Many western families were slaughtered by the Native Americans, and they scalped 600 men in a regiment in Kentucky, to which Proctor was marching *(pg 61, #1). There were many others who were killed either by scalping or by fire (pg 208, #4). Once Harrison was successful in recapturing Detroit and Lake Erie for the United States in 1813, the British ceased to dominate the war in the Midwest. This was mostly due to the alliances newly formed between the Native Americans and the Americans (pg 76, #1). Harrison's victory also ended British control of the territory that would become the American states of Minnesota, Michigan, and Wisconsin (pg 267, #9).

The Americans continued their assault on the division of the British Army in upper Canada, and they won the battle of the Thames, even killing Tecumseh. The Native Americans were paralyzed by his death (pg 111, #3). Later, American forces won the battle of Ft. George on July 5th, 1814 (pg 84, #1). This improvement in the war effort came about when members of separate militias joined the regular U.S. army and received training from Generals Winfield Scott and George Izard (pg 129, #5). But other problems would develop for American citizens during this time, and these were not caused by the British.

American citizens were constantly being attacked in Florida. This is where the future President of the United States, General Andrew Jackson, became a valued U.S. military leader. Jackson was given command of the New Orleans expedition, serving under General James Wilkerson. Jackson hated Wilkerson (pg 272, #4). Wilkerson split up Jackson's forces, who were then sent to Tennessee. Jackson walked home with his men to Tennessee in union with his troops. Jackson was known as a strict discipline enforcer and a fierce warrior, but he was also kind and loyal to the soldiers serving under him (pg 273, #4).

The Creek tribe was causing problems for U.S. citizens in Florida. The Creek had ties to Montezuma and the Aztec tribe (pg 277, #4). They had buildings erected of public interest, and had previously acted freely with the white settlers. Yet when Chief Red Eagle took control of the Creek, he was very upset that white families were multiplying. The Creek tribes were split into two factions: the upper Creek wanted war with the whites, and the lower ones did not. The upper Creek decided to attack the relatively unprotected Ft. Mims in August 1813, which resulted in the tragic massacre of 250 people, mostly women and children.

Outraged by this, Jackson wanted to attack the Creek nation and declare war against all southern tribes. Madison granted that request. Jackson's aggression towards the Native Americans stemmed back to the Revolutionary War, when he was a captive of the British. There, Jackson watched the Native Americans, on British orders, mercilessly attack other Americans. Jackson was treated harshly and unjustly as a British P.O.W., and he watched his mother and brother die as a result of it. This left Jackson an orphan at age 15 (pg 19, #2).

When Jackson opened his vicious assault against the Creek, he did have some help from other Creek tribe members. When the two sides met in battle, on November 3rd, 1813, Jackson's forces slaughtered the Creek severely. There were 186 Creek fatalities in this battle, compared to five Americans dead and 41 wounded. Jackson then met the Red Stick warriors at Talladega, where Jackson's forces killed 300 of the 1,680 Red Stick soldiers they faced (pg 66, #2). Jackson's mission against the Creek warriors had but one purpose, and that was the destruction of the Creek nation as a potential threat to the United States (for example, the Ft. Mims massacre).

On March 14th, 1814, Jackson met the Red Stick warriors again with his 5,000 troops at the Battle of Horseshoe Bend. Jackson enlisted the help of some of the Cherokee and Creek to fight with him against the enemy (pg 76, #2). In the end, Jackson and his forces annihilated the Red Sticks, killing 557 in battle, a number that later grew to 850. This is compared to 26 U.S. soldiers killed. As far as Jackson was concerned, the Battle of Horseshoe Bend ended the Creek War.

The Creek had severely misjudged the Americans' strength militarily, and they had no cannons (pg 284, #4). Jackson destroyed the Creek nation by slicing through it to Mobile, Alabama (pg 285, #4). In total, over 3,000 Creek warriors were killed in battle in the Creek War, which was approximately 15% of their total population. When the battle was over, Jackson did his customary burning of the entire town (pg 80, #4). Jackson continued his attacks until the entire Red Stick nation had no choice but to surrender. The conquered Native Americans referred to Jackson as "Sharp Knife" (pg 81, #4). During the peace negotiations, the Creek surrendered 23 million acres of land (pg 128, #3) by regretfully signing the Treaty of Ft. Jackson (pg 279, #9).

Madison stayed strong during this highly stressful and unpredictable period in American history. Russia had requested to act as a mediator for peace, but was rejected by President Madison (pg 554, #6). On May 24th, 1814, Madison predicted that the British army would try to destroy the city of Washington, and they did, in fact, do this. Yet before this act, the

United States successfully won the Battle of Chippewa in July 1814. Also, Major General Jacob Brown, who took over for the relieved commander General Wilkerson (who was not productive), repulsed British General Provost at Sackett's Harbor (pg 279, #9). American and Britain troops met with equal forces against each other at the Ft. George battle, and the United States triumphed. Never again would America lose a battle to Britain.

Sensing the loss of momentum in the war, the British decided to strike the U.S. capital of Washington. This expedition was led by British Admiral Alexander Cochrane, who had originally wanted to attack Baltimore. But after he met with British officials, the British changed their minds and attacked Washington (pg 319, #4). Baltimore offered one of the greatest sea ports in the new world, and was home to hundreds of American privateers. Yet the destruction of the capital, home to the president and Congress, would be a more lucrative prize (pg 319, #4).

After the British Navy successfully destroyed Commodore Matthew Barry's flotilla in the Chesapeake Bay, they moved up the Patoxent River towards Benedict, Maryland. The British stopped at Upper Marlboro, Maryland, where Ross stayed with Dr. William Beanes, who was a known British sympathizer. Beanes would later be arrested by Ross on the charge of jailing stragglers from British forces. Francis Scott Key would be sent to secure his freedom. After this battle, President Madison dismissed his Secretary of War John Armstrong and replaced him with Secretary of State James Monroe, who would hold both titles through the end of the war. Monroe would succeed Madison as president in 1817.

American General William Winder, who was commander of the 10th District in charge of protecting D.C., Maryland, and Virginia, was unable to track precise British movements, which forced him to divide his forces. British forces successfully routed much larger American forces at Bladensburg, which was blamed on poor American leadership that included President Madison and Secretary Monroe. As the battle was being waged, First Lady Dolley Madison saved as many items as she could from the Executive Mansion before fleeing to the Bank of Maryland, where they could be safely protected. Dolley Madison successfully saved cabinet papers, the White House silver, books, some personal belongings, and the Gilbert Stuart portrait of George Washington that still hangs in the East Room today.

British General Robert Ross ordered his troops to burn all government buildings, including the White House, the Capitol building, and the Library of Congress. Ross also ordered the burning of a private, non-governmental building, which was the headquarters of the newspaper *The National Intelligencer*, which was known as extremely pro-war and anti-British. American naval clerk Mordecai Booth was ordered to burn all ships under construction at the Navy Yard, in order to prevent the British from capturing them. However, before the British burned the White

House to the ground, they ate a full state dinner that Dolley Madison and the mansion staff had set out. The only government building that was spared by the British was the patent office, whose head had successfully argued to the British troops that its contents were private belongings and not governmental property.

The British army had to flee Washington the next day after a freak storm had struck. There are some historians who believe it was a hurricane, while some others believe it was a tornado. There are no other records of it hitting anywhere else but the Washington, D.C. area. When the British Army left D.C. due to the storm, they marched north to Baltimore, where they faced two defeats. One was at the Battle of North Point, where General Ross was killed, and the other at Ft. McHenry (#9).

These stressful events caused Madison to be on the verge of a nervous breakdown, yet he vowed to continue the fight until victory was achieved. After the burning of D.C., the war in America had assumed a new identity, because the citizens of America then strongly united to fight and defeat Britain. War fever seized cities like Boston, New York, Philadelphia, and Charleston (pg 328, #4). The U.S. Navy captured 165 British ships with only 22 of their warships. This was considered a sizeable naval victory.

After Britain's assault on Washington, it then chose to attack Baltimore, but the prior attack on Washington gave the United States time to prepare its defenses in Baltimore. On September 13th, 1814, the British bombed Ft. McHenry in Baltimore for 25 straight hours, yet the United States would not surrender. Francis Scott Key, who witnessed the bombing, wrote the "Star-Spangled Banner" during part of this assault on Ft. McHenry.

When the war first began, Key did not agree with it. The British ransacking of Washington, however, changed his mind (pg 343, #4). For example, the line "rocket's red glare, the bombs bursting in air" was a direct reference to Ft. McHenry (pg 344, #4). But the song was not adopted as the U.S. national anthem until March 3rd, 1931 (pg 344, #4).

In an effort to block Spanish help to the British, General Jackson invaded Pensacola on November 7th, 1814. This was to prevent the Britain from invading New Orleans through the Gulf of Mexico, and it was a successful military victory for Jackson. England's economy was being adversely affected by the war, and it was very unpopular with English citizens. The Tories in England wanted more U.S. territory, but their own people said no. By the fall of 1814, the British were prepared to invade New Orleans in a last attempt for control of the Mississippi River. Consequently, Jackson had 4,000 of his troops entrenched in Mobile, Alabama, waiting for the British attack.

England, which grew weary of the war, signed the Treaty of Ghent on December 24th, 1814, which gave America their coveted shipping rights and the right to be unmolested in the future. Yet it would be the Battle of New Orleans, led by Jackson and his forces, that would prove to be the final decisive battle of the war. This was because neither Britain nor America had ratified the treaty by January 1815. English General Edward Pakenham was sent to fight the Americans in New Orleans. This was the battle that became known as the Battle of New Orleans, and is one of the most significantly important battles in American military history.

The British had the disadvantage of attacking unfamiliar territory, against three lines of Jackson's troops in fortified positions, and at night. The United States, however, could and did conduct night raids against the British for precisely the opposite reasons. The American firepower was devastating and unrelenting. The British army never properly recovered from the New Orleans battle, sustaining more than 2,000 casualties and 46 deaths, compared to much lighter casualties for America (24 dead, 115 wounded, and 74 captured) (pgs 306-07, #8). The American victories at Plattsburgh and Lake Champlain were also deciding factors for the British to give up and formally ratify the Treaty of Ghent on February 17th, 1815, which officially ended the war and hostilities with Britain.

Four positive changes that the United States enjoyed from winning this war were:

1. Intersea commerce returned to normal following Britain and France's respect for American shipping.

2 The "Star-Spangled Banner," by Francis Scott Key, was written.

3. The Monroe Doctrine was established, preventing any European nation from trying to dominate anyone in North America.

4. A permanent army and navy were established.

Four more heroes from this war would eventually emerge as presidents: William Henry Harrison (hero of Tippecanoe), John Tyler (captain of the Richmond Brigade), Zachary Taylor (hero of Ft. Harrison), and Andrew Jackson (hero of New Orleans).

Bibliography

1. *The War of 1812*, by Henry Adams (Cooper Square Press, 1999), disc 1.
2. *The Life of Andrew Jackson*, by Robert V. Remini (Penguin Books, 1988), disc 2.
3. *War of 1812: A Compact History*, by Major James Ripley Jacobs and Glenn Tucker (Hawthorn Books, 1969).
4. *From Sea to Shining Sea*, by Robert Leckie (Harper Perennial, 1994).
5. *James Madison*, by Garry Wills (Henry Holt, 2002), disc 5.
6. *The Fourth President: A Life of James Madison*, by Irving Brant (Bobbs Merrill, 1970), disc 6.
7. *Andrew Jackson: His Life and Times*, by H.W. Brands (Random House, 2005), disc 10.
8. *1812: The War That Forged a Nation*, by Walter R. Borneman (Harper Collins, 2004).
9. *August 24, 1814: Washington in Flames*, by Carole L. Herrick (Higher Education Publications, Inc., 2005).

Chapter 5
The First and Second Seminole Wars, the Black Hawk War, and the Other U.S. Conflicts With Native Americans

To understand the problems that existed—and continue to exist—for Native Americans, it is necessary to examine the culture of the Native American, starting when Christopher Columbus landed in the New World in 1492. It was during 1492 that Columbus and the crew of his three ships had mistakenly thought that they had landed in India. Hence, the error of calling the native people Indians began. When Columbus landed in America, there were approximately 30 million native people living in Mexico and the Caribbean Islands. In addition to this, there were an estimated 70 million Native Americans in all of North America.

To use the technical appellation of Indians for these native people is factually incorrect, because Columbus and his crew were not in India, as he had originally thought. In fact, Columbus' discovery of the New World was really an invasion, and he was guilty of selling the native population into slavery, as well as of the torture and murder of many natives. Columbus and his crew committed these evil crimes against the Native Americans because, like many Europeans at the time, they were consumed with conquering new, undiscovered land. Another reason for the migration was because the land was unclaimed by a foreign entity and immediately available for colonization. Also, a huge dilemma for the native people derived from the diseases and sicknesses that the Europeans brought with them to the New World, which resulted in the deaths of many Native Americans.

In fact, because Columbus was so cruel to the Native American people, he was arrested and tried in Spain for these crimes. By 1506, Columbus died in Spain as a somewhat disgraced man. Yet the Spanish defeated the numerically superior Aztecs in 1521 (pg 81, #4). Whereas the Spanish were horribly cruel to the Native Americans, the French settlers had developed quite good relations with the natives. Then, English settlers arrived and were set to colonize the New World.

When the English landed in 1607, they named the place in Virginia where they landed Jamestown, after their King at home, King James I. The tragedy of this colonization was that the new English settlers brought with them the disease smallpox, which literally wiped out entire tribes of natives. In fact, the Pequot War in 1637 almost wiped out the entire tribe. The natives were faced with the choice of siding either with the French or with the English colonists in North America, because they were the two dominant European powers in North America. These two nations had been battling each other for years on the savage war scene in Europe (pg 124, #4). These hostilities between France and England continued into the 18th century, including King George's War, which lasted from 1740 to 1748 (pg 176, #4).

These hostilities would be revived during the years 1756-63, when a war for dominance in America existed, called either the French-Indian War or the Seven Years' War. The English army proved victorious in this endeavor, and this would cost France most of its colonial empire in America. This included the Canadian city of Montreal. During this war, the French military and the Native Americans formed an alliance to defeat the British military, because the British were enemies of the Native Americans and were fighting a guerilla war on their soil. Yet during this war, George Washington was successful in negotiating the help of Cherokee and Catawba warriors. This was established by the signing of the Treaty of Paris that officially ended the war.

The English were then in control of the majority of land in North America, and they desired to obtain the rich, fertile lands that the natives controlled. To the whites, land meant wealth, freedom, and power. To the natives, the land belonged to everyone and could not be sold, because the Great Spirit created it. This clash of beliefs led to the bloodshed and war that often occurred between the white, European settlers and the Native Americans.

The problems for the Native Americans started with the lack of a central government to effectively represent them in negotiations with the white settlers (pg 118, #2). Yet the natives did have an alliance with each other, called the Five Civilized Tribes, which consisted of the Chickasaw, Cherokee, Creek, Choctaw, and Seminole tribes.

In fact, the term "Seminole" in its native language means "frontiersman" (pg 13, #2). By around the time of the War of 1812, these aforementioned Five Civilized Tribes were led by a Shawnee chief named Tecumseh. Tecumseh was a notorious anti-white warrior who believed that the natives had a right to launch a massive assault on the white European settlers, in order to reclaim their homeland. These resentments had accumulated, going back to when President Washington and his Secretary of War, Henry Knox, attempted to sign the Treaty of Holston by offering to pay for land that was illegally obtained by the whites. Though Washington did want to live in peace with the natives, he and his army, during the Revolutionary War, attacked the Iroquois in Pennsylvania because they had previously attacked frontiersmen in northern Pennsylvania and had killed 200 settlers. When Washington's forces struck, they inflicted a large blow to the Six Nations alliance of Native Americans, from which they never fully recovered.

This trend continued when the next president, John Adams, signed the Treaty of Tellico on October 2nd, 1798, which gave the white settlers titles to lands that were in dispute. Adams' successor as president was Thomas Jefferson, who had bought the territory of Louisiana from Napoleon for $15 million. This land purchase added significantly more territory for the United States, and almost doubled its size. In both Adams' and Jefferson's administrations, they were successful in convincing the native people to trade their land in treaties to America, because they had to pay their outstanding debts to the white settlers. These debts were the result of the white settlers forcing their products on the natives, including alcohol, tobacco, and other goods (pg 49, #2). In fact, in 1871, Senator Eugene Casserly would say of the Native American, "I know what the misfortune of the tribes is. Their misfortune is not that they are a dwindling race, nor that they are a weak race. Their misfortune is that they hold great bodies of rich lands" (pg 222, #4).

By the time James Madison became president in March 1809, tensions between the natives and the European settlers had hit a boiling point. The aforementioned Tecumseh's father was killed by white hunters when Tecumseh was a boy. This act fueled his hatred and aggression towards white people. Tecumseh would later learn the art of guerilla warfare, which would make him famous. He would lead the awful Ft. Mims massacre on August 30th, 1813, that mercilessly killed between 250 and 275 white settlers, mostly women and children.

This evil massacre at Ft. Mims occurred during the War of 1812, so it was President Madison who recognized that this brutal act would begin the course of events that led our nation into what would be called the Creek War. In order to suppress this aggression, President Madison sent General Andrew Jackson to suppress the Creek warriors. As a result of this appointment, General Jackson declared war against all southern tribes in the United States. Jackson had grown up in a hostile Nashville, Tennessee, in which the natives and white settlers were killing each other regularly due to disputes over land that the natives considered to belong to them (pg 26, #2). In fact, according to the book *Andrew Jackson and His Indian Wars*, by Robert Remini, "To Jackson, killing Indians and driving them farther south and west was the only way to safeguard the Tennessee frontier. It was a necessary function of life in the Wilderness" (pg 27, #2).

Jackson began his campaign to attack the Creek by enlisting help from some other Creek tribes. However, the hostile Creek and Jackson's forces met on November 3rd, 1813, and the Creek were mostly slaughtered. There would be a second battle versus the Red Stick tribe in Talladega, and Jackson's 2,000 troops killed 300 of the 1,080 Red Stick warriors they met in battle. Jackson's reasoning for this battle against the Creek/Red Stick tribes was to destroy their potential threat to the United States and its citizens. An example of this threat was the massacre of Ft. Mims.

On March 14th, 1814, Jackson's 5,000 troops met the Red Stick warriors in what was called the Battle of Horseshoe Bend. Jackson's military men had the assistance of the Cherokee tribe and some Creek warriors at this battle. Together, they surrounded the Red Stick soldiers and launched a ferocious campaign against them. In fact, the carnage the U.S. forces inflicted on the Red Stick tribe was described as "hideous," with old men and five-year-old boys getting killed along with the Red Stick warriors. General Jackson's reason for this relentless campaign against the Creek/Red Stick tribes was that if they were allowed to continue on as a nation after the Ft. Mims massacre, they would have eventually allied themselves with Britain in order to establish a permanent Native American nation.

In total, at the Battle of Horseshoe Bend, 557 Red Stick warriors were initially reported killed in combat, but this number would grow to 850. In comparison, there were 26 U.S. fatalities in this battle. As far as General Jackson was concerned, the Battle of Horseshoe Bend ended the Creek War. There were over 3,000 Creek fatalities in this conflict, which comprised approximately 15% of their total population. The conquered Creek referred to Jackson as "Sharp Knife" (pg 81, #2).

When the Battle of Horseshoe Bend was officially over, Jackson commenced with his customary burning of entire towns in the surrounding areas (pg 80, #2). Jackson and his forces continued the military attacks on the entire Red Stick nation until they had no choice but to surrender. Because of Jackson's victory over the Creek, he was promoted to Brigadier General (pg 86, #2). The Creek/Red Stick War enriched the United States with 23 million acres of land in the south and in Florida. Jackson insisted that for America's security and survival, the Creek nation must be destroyed, and it was.

Days later, on November 7th, 1814, Jackson's successful military campaign at Pensacola cut the British off from invading the Gulf of Mexico region through Florida, which was England's original intention. Three thousand Native Americans had volunteered to fight for England during this time, and the U.S. government considered this an act of war by the natives. So, owing to the natives' association with England, Jackson chose to move all Native American tribes west of the Appalachian Mountains, for fear that the natives would become a security threat during a possible foreign invasion. Therefore, Jackson threatened the native tribes with violence, then offered them a bribe that they had little choice but to accept. The Native Americans never had a chance.

Accordingly, the Treaty of Ft. Jackson ceded all of the Creek territory to white settlers. After both the Creek War and the War of 1812, President Monroe, who took over for President Madison in March 1817, would eventually give permission to General Jackson to conquer Florida. But the Spanish and the Seminole tribe stood in their way. The British had also made allies among the Seminole tribe's population to cause problems with the settlers, despite England's loss to America during the War of 1812.

The Spanish military authorities had very little police presence there, and Native Americans attacked settlements all the time. On November 12th, 1817, the U.S. military burned down Fowltown, Georgia, in order to prevent attacks from there, as well as in response to hostile gunfire coming from the area. The Seminole tribe retaliated nine days later on November 21st, 1817, when they seized a boat on the Apalachicola River, which included seven women and four children, who were massacred, and six soldiers, four of whom were wounded. An outraged General Jackson told Secretary of War John Calhoun that Spain was required to keep the Native Americans under control, which it had failed to do (pg 134, #2). It was General Jackson who convinced President Monroe that something had to be done to stop them.

Accordingly, General Jackson was told to report as Commander of Ft. Scott in Georgia, where 2,700 Seminole threatened the region. President Monroe gave Jackson the authority, through Secretary of War Calhoun, to "adopt the necessary measures to terminate a conflict" (pg 134, #2). Although it may not have been the federal government's original intent to seize Florida outright, it was certainly General Jackson's. President Monroe then assigned Jackson to do exactly that. Jackson wanted to do it in 60 days.

On January 18th, 1818, Jackson left his home at Hermitage, Tennessee, with 4,000 troops and headed south to Florida. At the time, east and west Florida were split at the panhandle. General Jackson and his troops first went to Ft. Scott, Georgia, on March 9th, 1818. There, Jackson learned that the Native Americans were being encouraged by English agents to attack American settlements. This included two Englishmen named Alexander Arbuthnot and Robert Ambrister.

On General Jackson's birthday, March 15th, 1818, he and his troops reached Negro Fort in the panhandle of Florida, and renamed it Ft. Gadsden after it was rebuilt. On April 6th, 1818, Jackson's forces reached St. Marks in west Florida and demanded that the Spanish surrender to them immediately, which they did. The next battle was located at Pensacola on May 24th, 1818, where hundreds of Seminole warriors were stationed with substantial Spanish assistance to attack American settlements. They were encouraged to do so by two British agents located in Florida, namely the aforementioned Arbuthnot and Ambrister. When Jackson was informed of this, his temper exploded, and he and his forces struck with a vengeance at the Seminole opponents with howitzers and gunfire. When the battle was over, Jackson and his army easily claimed the city of Pensacola.

Arbuthnot and Ambrister were caught in Florida by the U.S. military, and Jackson returned to St. Marks in west Florida in a 100-mile march to hold court for these felons. In the trial, the two men were convicted of giving information detrimental to the U.S. military operation to the Seminole and Spanish, and were hanged on April 28th on Jackson's order for their crimes. This aggressive verdict would later get General Jackson into trouble because the two accused men had no counsel, and Jackson technically had no legal precedent to do this. Jackson hated the British partly because his whole family died in the Revolutionary War, leaving him an orphan at the age of 15.

Jackson's nemesis in the House, the Speaker Henry Clay, wanted the House to vote for a censure against Jackson. However, no action was taken against Jackson, and the House acquitted him on both counts (pg 168, #2). After the May 24th attack on Pensacola by Jackson and his forces, the Spanish governor retreated. By June 2nd, 1818, the Seminole War was over. There was even a discussion by the U.S. government of allowing General Jackson to claim Cuba, but because Jackson had become extremely ill, the idea was eventually scrapped.

As Robert Remini describes in his book *Andrew Jackson and His Indian Wars*, "There is no doubt that Jackson personally benefited to an extraordinary degree from the land wrenched from Native Americans" (pg 180, #2). For example, the land that was purchased from the Native Americans, which had extremely valuable soil to farm, was sold for $2 an acre. This was all land that used to belong to the natives. According to Jackson, the natives were a threat if they were employed to assist militarily during a foreign invasion.

As a result of these military victories, on February 22nd, 1819, Spain agreed to cede Florida to the United States for $5 million. This became official on October 24th, 1820. Because of Jackson's sheer military dominance in battle, Spain became convinced that the United States could not be stopped in its quest to seize Florida. By June 28th, 1818, Jackson returned to Hermitage a hero. However, President Monroe had some complaints from Spain and England afterwards regarding Jackson's actions in battle while conquering Florida. While Monroe told both of these countries that Jackson had gone too far, he was really in favor of Jackson capturing Florida. This paved the way for the Adams-Onis treaty of 1819, which Secretary of State and future President John Quincy Adams and his Spanish counterpart constructed, officially ceding Florida to the United States. This also led to the creation of the Monroe Doctrine in 1823, which stated that no European powers could interfere with colonization in North America, and which became the policy of America's foreign relations with the European community for the rest of the 19th century.

When Florida officially became a state in the Union in 1820, General Jackson was asked to be its governor, and he accepted. Yet Jackson would only accept the position as governor on a temporary basis, partially due to suffering from massive health problems. Nevertheless, Governor Jackson guaranteed equal rights for all free men and made rich and poor citizens more equal in society. Jackson resigned as governor after 11 weeks on the job on November 13ᵗʰ, 1821, satisfied with his performance and ready to return to Tennessee (pg 210, #2). Over a period of seven years (1813-1820), Jackson acquired for the United States millions of acres of land from the Native Americans, including territory in North Carolina, Florida, Georgia, Alabama, Mississippi, Tennessee, and Kentucky (pg 205, #2).

Jackson, as president, was obsessed with national security and somewhat of a racist, due to the cultural beliefs of the time. However, Representative Wilson Lumpkin of Georgia, who was a member of the House Committee on Indian Affairs, in an effort to push President Jackson's legislation on forced Native American migrations, said of the president, "No man entertains kinder feelings toward the Indian than Andrew Jackson" (pg 236, #2). The measure passed in the House by a 102-97 vote, thereby legislatively expelling the Native American from their homeland. At the end of President Jackson's eight years in office, 132 million acres of land were acquired, both east and west of the Mississippi River for $68 million. As president, Jackson also signed 76 Native American treaties, and established the Office of Indian Affairs (pg 242, #2).

The next Native American war with the United States was the Black Hawk War, which occurred when Chief Black Hawk and his people were forced to move from their home territory of Illinois to the unknown territory of Iowa. This war occurred in 1832, and both Abraham Lincoln and Jefferson Davis served for the Illinois militia (pg 247, #4). This war happened because while Black Hawk and his tribe spent the winter months in Iowa in 1832, he returned to find that white settlers had already moved into his homeland in Illinois. The whites had claimed that they had legally purchased the land from a land company. The Illinois militia, comprising 8,000 militia and 150 federal troops, went after Chief Black Hawk and his Sauk and Fox warriors into Wisconsin from Illinois. Black Hawk himself was a member of the Sauk tribe. The U.S. military forces were under the command of General Henry Atkinson.

Chief Black Hawk's forces were defeated after receiving cannon fire from a steamboat, which inflicted many casualties. When a federal force of 1,300 U.S. troops arrived to reinforce the troops already stationed to fight Chief Black Hawk's forces, the natives were trapped between two armies. Some native people were massacred even after they had waved the white flag of surrender. Chief Black Hawk was captured, along with his surviving troops, and was forced to surrender Native American lands located in eastern Iowa. This was done for retribution by the American government. As a result of this war, the Sauk tribe were forced from their homeland in Illinois, and were not allowed to continue fishing, hunting, planting, or even residing in the Illinois area. The Black Hawk War lasted three months, from May through July 1832. Black Hawk was captured in August. To make matters worse, the Bad Axe Massacre, which occurred on August 3rd, 1832, took place when the remaining Sauk and Fox tribes were massacred trying to cross the Mississippi into Iowa.

During this time, the massive waves of Native American removals began in 1830, when the Treaty of Dancing Rabbit Creek was signed by the Choctaw tribe of natives, who were generally considered friendly and cooperative with the American people. However, they became subject to the first relocation under the newly passed legislation of the Indian Removal Act of 1830. Under a mixture of bribes, whiskey, and broken promises made to the native people by the federal government, the treaty provided for the educational care of the Choctaw people by stating, "The U.S. agrees and stipulates that for the benefit and advantage of the Choctaw people, and to improve their condition, there shall be educated under the direction of the President and at the expense of the U.S., forty Choctaw youths for twenty years." This agreement was in exchange for the removal of all Choctaw tribes, through blackmail and agreements, and the relinquishment of 10,423,130 acres of land east of the Mississippi. The Choctaw moved to western Arkansas, which is now Oklahoma (pg 249, #2). The U.S. Senate ratified this on February 25th, 1831 (pg 249, #2). Of the 13,000 Choctaw who migrated on this trip, 4,000 died of hunger, disease, or overexposure to the sun. The natives were abused and mistreated during this forced migration.

Next came the Second Seminole War, which lasted from 1835 to 1842 (pg 250, #4). This war cost the U.S. government over $50 million, and is regarded as the least successful and most unpopular war in America until the Vietnam campaign over a century later. This war was provoked by the Seminole tribes in the Florida region resisting the forced removal of their people to the west, and fighting the American military to prevent this. The Seminole tribes were led by Osceola, who was considered the best and most capable leader in Seminole culture. American forces deployed in response to this hostility from the Seminole are estimated to be approximately 5,000 to 10,000 (pg 251, #4).

The war began when Osceola and company attacked the U.S. Indian agent who was attempting to get the Seminole to adhere to the Indian Removal Act of 1830. When this occurred, the American troops struck back at the Seminole, capturing Osceola that year by tricking him, under the terms of a white flag, to a peace council and arresting him. He died in prison shortly thereafter. He would be replaced by Billy Bowlegs. Bowlegs led attacks against a government trading post operating on Seminole land, and killed many American troops stationed there. The Seminole retreated into the Everglades during the day and raided the Americans at night. The

American army was forced to pursue the enemy into the swampland, which subjected the troops to bouts of malaria. Winfield Scott and Edmund Pendleton Gaines were the two commanding generals for this military campaign. President Jackson was blamed for this unpopular war, which occurred on his watch as commander-in-chief.

During his last year as president, Jackson ordered the removal of the Cherokee tribe's 16,000 members from their homes in Georgia, during the middle of winter, and forced their migration to Oklahoma under the supervision of seven thousand federal troops. This forced migration of the Cherokee people, in which one out of four died on the journey alone, was forever known in the darkness of American history as "The Trail of Tears" or " The Trail Where They Cried" (pg 251, #4). This forced migration took the survivors six months to complete. The forced migrations continued in 1838, under the leadership of President Martin Van Buren. The reason for forcing the removal of the Chickasaw tribes from the Mississippi and Alabama territories was so that the land could be taken for cotton production.

Even after his presidency, and to his dying day (which was June 1st, 1845), Jackson genuinely believed that what he had accomplished literally rescued the Native Americans from inevitable annihilation. Jackson also thought that he saved the Five Civilized Nations from probable extinction (pg 281, #2). The reservations that the Native American people were forced to live on were set up in the 1840s to 1880s to literally separate the Native Americans from the white settlers. The next installment of the native/U.S. wars was the Apache War, from 1846 to 1868, which occurred in New Mexico when different Apache tribes combined forces with miners against U.S. army personnel who were invading Apache land. The Apache was led by Mangas Coloradas, who would battle the Americans for years in an unsuccessful campaign to keep the Americans from the New Mexico region.

The horrible Pomo massacre occurred in 1849-1850, when two Pomo men killed two abusive American ranchers, and American authorities took the situation too far, allowing soldiers to round up 130 Pomo men, women, and children, fishing by a nearby lake, and kill them all in a brutal act of hatred. The trouble continued in 1851, when Ft. Defiance was built in Navajo country in Arizona to keep the Navajo under control. They were previously under the loose control of the Mexican authorities, but were now forced to deal with the Americans for use of their own land and water. The fort was later attacked in 1860 by 1,000 Navajo warriors, and though the Navajo were driven off by the U.S. forces, the fort was abandoned a year later, due to their withdrawing to fight the Civil War. The Gadsden Purchase gave the United States the final pieces of territory of Arizona and New Mexico in 1853. That same year, the Omaha Indians ceded more than 43 million acres of land to America by signing the Missouri Treaty, while they only got to keep 300,000 acres (pg 261, #2).

Next came the Third Seminole War, which took place from 1855 to 1858 and involved Billy Bowlegs again attacking American government workers, who had vandalized Bowlegs' crops. He and his forces of Seminole attacked Americans at night, and retreated to the Everglades during the day. Though the Americans successfully beat the Seminole tribe into submission, it came at the financial cost of $40 to $60 million.

During the Civil War, members of the Five Civilized Tribes of Indian Territory (Oklahoma) had signed treaties with the Confederacy. Consequently, 6,435 Native American troops would serve for the Confederacy during the Civil War. The Native American believed that they had a better future in the Confederate states. Even Billy Bowlegs eventually ended up serving, but for the Union against the Confederacy However, in 1862-63, the Navajo-Apache Wars took place, in which General James Carleton of California was appointed head of the military department of New Mexico and was ordered to take Santa Fe back from Confederate influence and Apache/Navajo control. Carleton wanted the Native American tribes to be relocated to eastern New Mexico, close to Texas. Carleton had the famous Native American locator Kit Carson find the natives by destroying their means of survival. This included burning native crops, killing animals belonging to the natives, killing their supplies of sheep, and confiscating any valuables that the Navajo had. Their mission was to starve the Navajo/Apache into submission, and to shoot anyone who fled the U.S. forces. The Navajo/Apache forces eventually surrendered and were forced to march 350 miles from Fort Defiance to Bosque Redondo (pg 270, #4).

But tragically for the Native Americans, the worst was yet to come. In 1864, the Sand Creek Massacre took place, when white gold hunters willfully raided native land, led by a Methodist minister named John Chivington, who was running for Congress on a "kill the Indians" platform. He was intent on attacking the Cheyenne and Arapaho tribes located at Sand Creek, who were led by a Native American named Black Kettle. They were known as being peaceful people. Chivington ordered 700 U.S. troops to round up these peaceful Native Americans, who were attacked in an unprovoked manner, and on November 29th were murdered, execution style. These included elderly men, women, and children. This was a barbaric act on the part of the American troops, who had no right to systematically kill whomever they saw fit, and clearly overstepped their authority.

The Cheyenne and the Sioux tribes struck back hard against the white settlers, who were attempting to construct military forts in Sioux country but were successfully defeated by the Sioux tribes, led by Chief Red Cloud (pg 275, #4). However, these victories were not enough to let the natives keep their prized land, and they were forced to relocate to reservations of the white man's choosing.

Next came the Apache Wars from 1876 to 1886, in which the U.S. government wanted to relocate a tribe called the Chiricahua Apache to a reservation on the Gila River, in southeastern Arizona. The Apache leader, Geronimo, did not agree with the relocation, and he raided the white parties from Arizona to New Mexico, escaping each time. Finally, in 1886, American General George M. Crook was placed in charge of 5,000 troops, and he forced Geronimo to surrender in 1887 (pg 286, #4) Also, the Sioux resistance was finally broken in 1877, when the last free tribes of Sioux were relocated to reservations, and their leader, Crazy Horse, was killed while meeting with General Crook, who witnessed Crazy Horse bayoneted by another Native American disguised as a policeman. Native leader Sitting Bull fled to Canada, with his tribe of Hunkpapa Sioux, and refused to live on American reservations. However, these two native leaders (Sitting Bull and Crazy Horse) were present on June 25th, 1876, when a combination of Sioux and Cheyenne forces wiped out General George Custer's 7th Cavalry regiment troops, killing Custer.

Another of the conflicts, and one of the last between American troops and Native American warriors, was the battle at Nez Perce, which was in dispute after the fraudulent treaty that wrongly ceded six million acres of Native American lands to the white settlers. When Chief Joseph of the Native Americans was attempting to work out a peaceful solution with the Americans, the Americans shot a team of Nez Perce peace negotiators, who had a white flag with them. In retaliation, they began a war against the American troops that lasted for four months and spread out over 1,500 miles, territory that had successfully eluded the U.S. military's control, almost to the border of Canada. When the natives eventually surrendered, they were sent to Ft. Leavenworth, Kansas, and were kept as prisoners of war for eight years, before being released in 1885 (pg 288, #4).

In 1882, President Chester Arthur created 4,000 square miles of public land in Arizona, and made it official Indian reservation land for "the Moquii Indians, and such other Indians that the President should decide to settle thereon" (pg 292, #4). Geronimo was eventually captured by General Nelson Miles and 5,000 U.S. troops in the Sierra Madre mountains. Instead of being returned to the San Carlos reservation, President Grover Cleveland sent Geronimo and his Apache members to Ft. Marion in Florida, and ordered Geronimo executed by hanging. This did not happen, however, and Geronimo died in prison at Ft. Sill in 1909. The native children were ordered to be sent to Carlisle Indian School in Carlisle, Pennsylvania to be educated, and to never see their family members again.

In 1887, Senator Charles Dawes of Massachusetts passed the Dawes Act through Congress and obtained President Cleveland's signature; it said that white civilians could attain full title to reserved Indian land that Native Americans claimed belonged to them through tribal leadership.

This law took away, over a period of 47 years (1887-1934), over 86 million acres of allotted Indian land, and gave it to the whites. This act was done, in part, to protect the railroad interests of expanding via larger land acquisitions in the west. Another blow to the natives was President Benjamin Harrison's proclamation that white settlers could occupy lands in Oklahoma that were originally promised to the Native Americans. Fifty thousand whites claimed more than two million acres of Indian land, and the Native Americans were forced to obey the white man's law, or there would be war.

But perhaps the final blow to the Native Americans came in 1890, at the Massacre of Chief Big Foot's Sioux tribe at Wounded Knee in South Dakota. After Sitting Bull was killed earlier that year, Chief Big Foot had decided to move to the more protected Pine Ridge region. U.S. troops had been given the order, however, to eliminate the Plains Indian tribes in the region (pg 299, #4). When they had peacefully surrendered to the U.S. 7th Cavalry, a gun had accidentally gone off, which made the U.S. troops open fire with four howitzers against the surrendered natives killing more than 146 men, women, and children. This horrible massacre of innocent people ended the Plains Indian Wars, and the native population in America, once estimated in 1300 to be as high as 10 million (#4), was reduced to 237,196. By 1990, however, the population rose to 1,959,234 (pg 300, #4).

By the 1900s, the wars that led to the containment of the Native Americans on their reservations were over (pg 304, #4). Native land that was once estimated to be approximately 155 million acres in 1881 was reduced to 59 million acres in 1910 (pg 304, #4). The railroads had cut the west into three corridors, and the buffalo herds had completely disappeared. By 1928, the horrible poverty and suffering of the native peoples on reservations were brought to light by Congress, which wanted a comprehensive study on the native people's conditions in society authorized, because Native Americans had been relocated to some of the worst lands that no white man wanted. Consequently, in 1934, the Meriam Commission was created to help reform life for the native people. For example, because Indian reservations had no self-sustaining economies, Congress passed the Indian Reorganization Act, which recognized the failures that were inflicted on the native people by the allotment policy and the poor conditions on Indian reservations. The Indian Reorganization Act recommended the reorganization of tribal governments to help revive reservation economies with special projects: building roads, water facilities, and small industries (pg 304, #4).

This was true during the Roosevelt dam construction process in 1912, in which Apache workers built the apparatus. More than 10,000 native men enlisted for service with the Allied Powers for America. This gave the Native Americans more opportunity for job growth. Also, in 1919, all Native Americans who had served in the military for America during World War One were granted full citizenship of the United States by

Congress. This led to the 1924 Indian Citizenship Act that granted all Native Americans full U.S. citizenship (pg 328, #4). This gave the Native Americans the right to vote in elections.

Later, in World War Two, all native men were required to enlist in the national draft, although many states still did not let native people vote in 1941. However, 24,521 native men did serve for the United States in World War Two, including the regiment that was captured by the Japanese in the Philippines and forced to partake in the infamous Bataan Death March. The famous photograph with the U.S. Marines raising the U.S. flag at Iwo Jima included Native American soldiers. In total, more than 70,000 native men and women either served in the military or in the defense industries (pg 342, #4). Ironically in 1996, the white residents of Wallowa Valley, Oregon, had invited the members of Chief Joseph and his Nez Perce people back to live, and are in the process of building a memorial to the late Chief where he is buried.

In conclusion to this difficult-to-write about period in American history, it is clear that the Native American people were wronged by the European settlers over the centuries by being robbed of their homeland, culture, and most of their very lives. It is also clear that much more now needs to be done to reconcile with the native people, and to give them more opportunities in our society, including a stronger voice in our government. The history of the natives should always be preserved, and their descendants should be included and welcomed into the melting pot that makes up American society. Though we cannot change the violent and repressive past, all we can do now is to make the Native American members of our population as productive and as prosperous as the American dream allows, and enable the native people to live as peaceful and as happy a life as they can lead. The National Museum of the American Indian in Washington, D.C. is a great illustration of the Native American heritage and prestige, and well represents the native voice in not only American culture, but also that of the world. I hope that the native people and the rest of the American population can live in peace and prosperity in congruence with the American dream.

Bibliography

1. *Lincoln: A Contemporary Portrait*, edited by Allan Nevins and Irving Stone (Doubleday, 1962), disc 1.
2. *Andrew Jackson and His Indian Wars*, by Robert V. Remini (Viking Books, 2001), disc 6.
3. *The Life of Andrew Jackson*, by Robert Remini (Harper Perennial, 2011).
4. *Native American History: A Chronology of a Culture's Vast Achievements and Their Links to World Events*, by Judith Nies (Random House, 1996), disc 10.

5. *The Mammoth Book of Native Americans: The Story of America's Original Inhabitants in All Its Beauty, Magic Truth, and Tragedy*, edited by Jon E. Lewis (Carroll and Graf, 2004).

6. *Andrew Jackson: His Life and Times*, by H.W. Brands (Anchor, 2006).

7. *Adams*, by David McCullough (Simon & Schuster, 2001), disc 5.

8. *The Wars of America*, by Robert Leckie (Harper Perennial, 1993).

INTERVIEW WITH DR. ROBERT KEIGHTON

Thomas Athridge: How much influence in Jackson's aggressive demeanor did the lack of a father have, as a result of Andrew Jackson, Sr.'s death the same year Andrew was born?

Robert Keighton: He was extremely aggressive, even as a young man. Whether this was caused by a lack of a father, his early experiences (with the British), or his innate personality is difficult to say.

TA: How traumatic was Jackson's Revolutionary War experience, including being taken prisoner by the British in South Carolina, which resulted in the deaths of his brother Hugh, in battle, and his beloved mother, Elizabeth, by disease?

RK: I am sure that this had a significant impact on his life and character. He lost 2 brothers: Hugh apparently died of natural causes while fighting, and Robert died after being held prisoner by the British (I believe of smallpox). His mother, who died of cholera after nursing American prisoners in Charlestown, South Carolina, was an enormous force in Andrew's life.

TA; How strong was the allegiance between the Spanish and Seminole Indian fighters who fought Jackson's troops at Pensacola?

RK: The Seminole fled into Spanish Florida after conflicts between the settlers and the Indians. I do not know how strong the bond was between the Spanish and the Seminole, but they had a common interest. The conflict with Jackson's troops eventually led to the sale of Spanish Florida to the United States in 1819.

TA: Did Jackson truly want to conquer and occupy Cuba, or was he just too ill?

RK: I assume you are referring to the time when Jackson was territorial governor of Florida and his relations with the outgoing Spanish governor, which I am not sure of.

TA: Did Jackson obtain a Napoleonic image as an indestructible military leader?

RK: Many certainly viewed him in this light.

TA: How frightened were the Spanish of Jackson, when Spain ceded all territory east of the Mississippi River to the United States?

RK: I believe that the Spanish always had a healthy respect for Jackson's prowess.

TA: How affected was Jackson emotionally by the death of his wife, Rachel, on December 28th, 1828, right before his inauguration as president?

RK: This had a profound effect on him. He was devoted to her and was beside himself with grief when she died (for a while, he did not believe she was even dead).

TA: How strong was the tension between Martin Van Buren and Jackson's Vice-President, John Calhoun?

RK: These were two politically ambitious men, who clearly developed a tense relationship. The political winner was Van Buren, who became Jackson's successor as president.

TA: To what degree is Jackson personally responsible for the forced Native American evacuation west of the Mississippi River, and for causing the infamous "Trail of Tears" march, resulting in 4,000 Native American deaths?

RK: Jackson was not a supporter of Native American rights, and while I cannot answer the specific question, my guess is that he had a major responsibility. The Indian Removal Act was passed in 1830, and Jackson certainly supported the removal of the Native Americans.

TA: How much power did Jackson bring to the Executive office, in relation to veto power, and presidential relations between the House and Senate?

RK: Jackson was not afraid to exercise authority, including the veto, and tried his best to hold sway over the House and Senate.

TA: How did Jackson keep the state of South Carolina from seceding?

RK: By threats and bellicose talk, but also by an expression that he was willing to make concessions. This followed the Ordinance of Nullification adopted by South Carolina in 1832.

TA: How did Jackson get the government of Spain to agree to pay $600,000 for seizing American ships during South America's war for independence?

RK: I am not aware specifically how he did this, but Jackson's reputation was such that other governments paid attention to him.

TA: How influential was Jackson in acquiring Texas from Mexico?

RK: I am not able to answer this. He was ambiguous about the struggle for Texas' independence, for various reasons, but he did recognize Texas' independence the day before he left office.

TA; If Jackson were alive now and in power, based on his past history, how would he have handled the September 11th, 2001 terrorist attacks, and the current war on terror?

RK: He would have been extremely aggressive in his response.

Thomas P. Athridge

Chapter 6
The Mexican War (1846-48) and President James Polk's Decisions in Time of War

The war that existed between the United States and Mexico during the years 1846-48 was over the American vision of "Manifest Destiny" in wanting to wrest Texas, New Mexico, and California away from Mexican control and add them to the American union. The term "Manifest Destiny" was described as God's call on the American federal government to take land all the way to the Pacific Ocean. Mexico, at the time, was described as unstable and chaotic, partly due to its economy.

The Mexican economy then consisted of two categories of people: the rich (Gapuchines) and the poor (Creoles). Between April 1829 and December 1844, Mexico had 14 different leaders. During this time, Stephen F. Austin had encouraged white settlers to move to Texas and colonize the land (pg 409, #1). Texas had no tariff to pay, which meant that land that was purchased was especially cheap and affordable.

While President Andrew Jackson was serving his second term, the United States was prospering economically. It began with Jackson's fight against the Second Bank of the United States and its president, Nicholas Biddle. Jackson's pro-bank opponents, such as Henry Clay and the aforementioned Biddle, became a political party known as the Whigs, who had beliefs similar to British Whigs, who opposed monarchical parties and believed in strong executive leadership. This was very similar to Jackson's own views of the Office.

The struggle that Jackson had with the national bank could be compared to a conflict between democracy and aristocracy. Jackson's opponents insisted that he was responsible to the needs of Congress, and not the people. An angered Jackson responded, "the president is a direct representative to the American people," making him responsible to them first. In the end, Biddle and his pro-bank allies lost their argument to Congress, and Jackson prevailed by terminating the national bank and beginning charters on state banks.

Other Jackson financial victories as president included the Deposit Act of 1836 and the Coinage Act of 1834, which helped spur the economy into much-needed success. Jackson was successful in reopening trade with the British West Indies, in direct contrast with the John Quincy Adams

75

administration, which had poor relations with the British in trading rights and practice. Jackson did manage to restore good diplomatic ties with England, which also wanted improved relations with America.

Jackson showed diplomatic and economic brilliance in negotiating with foreign governments to pay past debts for services rendered by the U.S. government. For example, Jackson got Spain to pay $600,000 for illegally seizing U.S. ships during South America's wars for independence. Jackson was also successful in convincing Denmark to pay $650,000 for detaining American ships from 1808 to 1811. Jackson even convinced the French, after heated disagreements, to pay its debt owed to the United States. This result was almost the end of relations between France and the United States, and only came about with the assistance of English negotiators. Once the agreement between the United States and France was settled, relations between France and America were ultimately improved.

Because of the shrewd political tactics by Jackson, he succeeded in the unimaginable task of paying off the national debt in 1835. Jackson also accomplished the quarterly distribution of a surplus of money in excess of $5 million to U.S. states throughout 1837. It was during this economically prosperous time in American history (1835-37) that President Jackson sought to acquire the territory of Texas for the United States.

At the time, Texas belonged to Mexico, which was very unwilling to part with this land. During both the presidencies of John Quincy Adams and Jackson, the United States made financial offers to the Mexican government to purchase Texas, but were refused (pg 322, #4). The relationship between Mexico and America was abysmal at the time.

Among the problems that existed between the two nations was that white Texas settlers were bringing slaves with them to Texas, which was in direct contrast to Mexico's anti-slavery policy. The Mexicans were Catholic. As a result, the Texas revolution began in 1835, in which Texans, led by Sam Houston, fought the Mexicans for their autonomy. They did not necessarily do this, however, to join the United States.

A big battle that occurred during this time was the Battle of the Alamo, which concluded in March 1836, when 200 Texas civilians, led by Lieutenant Colonel William Barret Travis, perished defending the Alamo against the Mexicans. However, these murdered Americans fought ferociously in battle, and took 1,544 Mexican soldiers with them into death. Hence the phrase, "Remember the Alamo!"

Another bad incident between Mexico and the United States occurred at Goliad, when a man named James Fannin and his volunteers from Georgia, Alabama, Louisiana, and Arkansas were attacked and surrendered to the Mexicans, with the understanding that they would be treated as prisoners of war. Instead, over 300 prisoners were shot and killed by Mexican forces. As a result of these atrocities, the United States battled and defeated Mexican forces at San Jacinto in 1836 (pg 19, #2).

This really upset Mexico, which was also angered that U.S. forces massacred the Mexicans to a more extreme extent than the Mexicans had to Americans at Alamo and Goliad together (pg 477,-#4).

Jackson's successor to the presidency, Martin Van Buren, was defeated in the election of 1840 by the Whig candidate, and former hero of the War of 1812, William Henry Harrison. He became the 9th President of the United States. Van Buren's presidency was hurt by the economic panic of 1837, which financially struck the middle and lower classes (pg 59, #4). However, Harrison would fall extremely ill from the unusually long inaugural address that he gave in bad weather, and from not wearing a proper jacket. He shockingly died 31 days into his presidency, on April 4, 1841. He was succeeded by John Tyler.

Tyler would only serve one term as president, and he was succeeded by James K. Polk. When Polk was a member of Congress, before his term as governor of Tennessee, he was known for being a strong Jackson supporter and very anti-Whig. Polk entered Congress at age 29 in 1824, which was the same year Jackson lost his presidential election to John Quincy Adams in a disputed election that was settled with the Adams-Clay deal behind closed doors, which Jackson proclaimed as a "corrupt bargain." Polk's illustrious career included him being a state legislator, a U.S. Congressman from Tennessee, then Speaker of the House, governor of Tennessee, and finally president (pg 2, #3). Because of Polk's friendship with Jackson, he would later be named "Young Hickory" (pg 3, #3).

Polk was born in Charlotte, North Carolina, on November 2nd, 1795 (pg 11, #3). He might have been considered odd for his time, because he followed no specific religion. In fact, Polk did not become a member of any religion until he was on his deathbed, at age 53. As a child, however, Polk was raised in a Presbyterian-dominated town. Therefore, as Polk grew up, religion would be second to politics in his life. Polk strongly believed in Jefferson's idea of separation of church and state.

As a teenager, Polk suffered from gall stones and took a chance on a very risky and painful surgery to correct his problem with the available technology. The surgery did relieve his gall stone problem, but it also left him sexually impotent and childless for the remainder of his life. After this medical ordeal, Polk concentrated on his studies in the private school system. It was there that Polk did exceedingly well.

All of Polk's academic work paid off, and he succeeded in getting into the University of North Carolina at Chapel Hill. Polk did this when he was only 16 years old (pg 21, #3). After his success there, Polk became a lawyer in 1820, and he was considered to be an excellent lawyer. In fact, Polk argued as the plaintiff's counsel in the case Williams v. Norris in front of the Supreme Court, which he won (pg 24, #3). Polk also joined the Masons and enlisted in his local militia, where he would receive the rank of colonel (pg 25, #3).

Polk married Sarah Childress on New Year's Day, 1824. Polk was heavily influenced by his mentor, Andrew Jackson, when he entered Congress. However, Jackson was known for having explosive tantrums and regularly getting into street fights or duels. In fact, during one duel, Jackson killed a Nashville citizen named Charles Dickinson. Jackson was also wounded in a street shooting with Thomas and Jesse Benton. Ironically, Jackson would later work with Thomas Benton when Jackson was president and Benton was a senator. But Polk, who won his House seat seven times, was credited with keeping a controlled temper on the House floor during heated debates, which normally often resulted in brawls or duels outside.

After serving seven terms in the House, Polk would successfully win the governor of Tennessee position in 1839. However, after failing to institute his state banking reform legislation, he lost his re-election bid in 1841. He also lost a third run for governor in 1843, to the same man he lost to in 1841. Polk also lost a chance to be the Democratic vice presidential candidate in 1840. Polk had to carry the political baggage of Van Buren's failed policies, which hurt him.

By the year 1844, Polk was attempting to position himself for a possible vice presidential run, with former President Van Buren as the Democratic nominee. However, political events drastically changed the scenario and instead elevated Polk to the presidency and left Van Buren out of the nomination (pg 69, #3).

Henry Clay, who lost the Whig nomination to William Henry Harrison in 1840, and former President Martin Van Buren, the supposed frontrunner for the Democratic nomination in 1844, were normally political opposites. Yet they actually teamed up in their mission to thwart Texas from joining the union and becoming a state. The reason for this was that Texas would have to be either a slave state or a free state, thus tipping the balance of slave (13) and free (13) states (pg 72, #3). Therefore, both Van Buren and Henry Clay did not want to give a state to either side of the slave question, so they jointly believed that no new state, like Texas, was needed. Polk, though politically allied to Van Buren, did disagree with both of them and believed that Texas should join the union. The general public seemed to agree with Polk.

Though this seemed to hurt Van Buren's chances for the nomination, it also hurt Clay in the South, whose citizens rejected Clay's wish to prevent slavery in Texas. While former President Jackson was in ill health, he saw that these events would not be enough for his former secretary of state or his vice president to win the election. Based on this assumption, Jackson lobbied hard to the Democrats for the nomination to go to Polk. At that time, no one even considered Polk as a presidential nominee.

On the first ballot, Van Buren was clearly the frontrunner, yet he did not have the votes needed for the two-thirds majority. Other candidates crowded the field and eroded Van Buren's momentum, such as Lewis Cass of Michigan, Rich Johnson of Kentucky, and James Buchanan of Pennsylvania. Polk was not even nominated until the eighth ballot, by New Hampshire. But by the ninth ballot, with Jackson's lobbying support for him, Polk had a commanding lead of 233 votes. Thus, he had the nomination. Polk came to be seen as the first "dark horse" candidate to be nominated for president, which was seen to be an attractive alternative to his Whig opponent, Henry Clay. Also helping Polk was his association with Andrew Jackson, and he inherited the position of annihilation of corruption by the federal government due to his kinship with "Old Hickory."

On the controversial subject of slavery, though, the abolitionist movement had strength in the North, whereas the South strongly defended its institutions on the basis of its economic needs. Therefore, to have campaigned for president on an anti-slavery platform would have been too controversial for the time. According to the author of *James K. Polk*, John Seigenthaler, "Historians generally agree that presidential aspirants such as Clay, Polk ,Jackson, Van Buren, and Buchanan had no choice but to tolerate the institution of slavery, for to stand openly for abolition was to invite defeat at the polls ... and risk civil war" (pg 85, #3).

Because of this potentially explosive issue, Polk campaigned on the platform that it would be up to the voters of Texas to decide. Polk also believed that to mercilessly beat a slave was an excellent means of enforcing discipline. This philosophy of evil was partly due to the incredibly racist culture of his time (pg 86, #3). This evil practice was exercised by both Clay and Polk, but a third-party candidate named James G. Birney from Michigan ran on a Liberty-Abolitionist ticket, which in turn took some of the abolition vote away from Clay.

Tariffs had become a hot-button issue for the campaign for the presidency in 1844. The northern states were in favor of protecting and implementing tariffs, whereas the southern states wanted to have fewer of them (pg 89, #3). When Polk's record as a House member was examined, it was discovered that he had campaigned against them in 1829. Sensing the potential political trouble from this issue, Polk made an effort to achieve a compromise with the North, and he assured northern Democrats that he would vigorously protect the tariffs as president (pg 90, #3). This enhanced Polk's overall popularity, and Clay's association with the Tyler Administration had caused him to be wounded politically. In those days, politicians did not campaign in the same way that politicians in the 21st century regularly do.

The Whigs campaigned on the platform that Polk was a nobody who had no political office, and asked the question, "Who is he, anyway?" (pg 92, #3). In response, Jackson returned the favor to Polk by campaigning for him in Tennessee and promoting Polk as someone who always stuck with Jackson during his turbulent and sometimes controversial life. Jackson's and Polk's close association helped to highlight Clay's and John

Quincy Adams' "corrupt bargain." This helped further damage Clay's credibility with voters. Polk sharpened his credibility with the American voter by saying that he would only run for one term, in order to prove that his service to the public as president would be his primary function. Adding to Jackson's credit in campaigning for Polk, he convinced sitting President Tyler not to run as a third-party candidate and to drop out of the presidential race. This helped Polk in the presidential race of 1844.

On election night, even though Clay carried most of the West and Polk's home state of North Carolina, it was the Democratic candidate Polk who won the 15 crucial states, compared to Clay's 11, to win the presidential election. Polk received 170 electoral votes to Clay's 105, making him the 11th President of the United States in the 1844 election.

Once Polk was inaugurated into the presidency on March 4th 1845, he would be described as an obsessive workaholic and a strict perfectionist (pg 103, #3). Even though Polk picked James Buchanan to be his secretary of state, Polk often acted as his own secretary of state. From the first day of his term, Polk sought to obtain the vast, northwest portion of our country for settlement, which was disputed by England (pg 123, #3).

Meanwhile, Texas had grown to 50,000 settlers, versus the 7 million Mexican people living as their neighbors. In Polk's vision of Manifest Destiny, he sought to acquire Texas, California, and New Mexico from Mexico. By this point, Polk had successfully avoided entanglement with England by agreeing to give Britain possession of lands above the 49th parallel, and the United States would gain the land that would become the states of Washington, Oregon, and parts of Wyoming (pg 128, #3). Polk was not, however, looking to militarily seize land from Mexico; rather, he was looking to purchase it. Polk offered to pay the Mexican government $15 million to $40 million for this acquisition (pg 519, #1). However, this proposal was angrily rejected by the Mexican president, José Joaquín de Herrera, in 1845.

To add to the tension between the United States and Mexico, Texas officially became a U.S. state on December 29th, 1845, during Polk's first year in office. This caused Mexico and the United States to be at each other's throats, and caused diplomatic strain between the two (pg 135, #3). The ensuing war between the United States and Mexico could be blamed on both Mexican political instability and failure of diplomacy (pg 20, #2). By March 28th, 1846, Mexico had broken all diplomatic ties with the United States. The U.S. government declared that the border of the United States and Mexico was at the Rio Grande River, while Mexico claimed that the border was farther north, at the Nueces River. Mexican troops were commanded by Mariano Arista (pg 43, #6).

Polk had sent General Zachary Taylor and 3,500 of his troops to the region. In a last-ditch effort to avoid a hostile conflict, Polk sent John Slidell to the Mexican government to offer a last financial settlement for the peaceful acquisition of Texas up to the Rio Grande border, which the

Mexican government had steadfastly refused. As a consequence of this failure, Polk sent orders to General Taylor to go to the Rio Grande border and protect the border regions. General Taylor had earlier achieved high praise and military fame for his actions at Ft. Harrison during the War of 1812 that led to victory against the British. Taylor's nickname was "Old Rough and Ready." From a political point of view, however, Taylor and Polk were the complete opposites of each other. Taylor was a Whig, and Polk was a Democrat.

Accordingly, Polk was very harsh in his criticism of Taylor. Polk's other military commander was General Winfield Scott, who was also a Whig, and both Scott and Taylor did well in previous military campaigns. Polk did have a friendship with Sam Houston, who was then president of an independent Texas. Polk wanted to go to Congress to ask for a declaration of war versus Mexico, but both Polk and Congress believed that Mexico should fire the first shot. The Mexican army did just that on April 24th, 1846, aggressively killing 11 of Taylor's soldiers and taking prisoners. This occurred on the "Texas side" of the Rio Grande. Polk then went to Congress to formally declare war on Mexico on May 13th, 1846, with a House vote in favor of war by 174 to 14 and a Senate vote in favor of war by 40 to 2. Polk signed the law, officially going to war with Mexico.

At the start of the war, the U.S. army had less than 7,000 enlisted men serving (pg 24, #2). Plus, on May 9th, 1846, U.S. and Mexican forces, led by General Arista for Mexico, clashed at Point Isabel, which would leave48 U.S. soldiers dead and 128 wounded in this fierce battle. However, this grim statistic was only one sixth of Mexico's troop losses. Polk, in an effort to install a pro-U.S. ally in the unstable position of Mexican leader, chose exiled former Mexican leader Santa Anna to replace Mariano Paredes y Arrillaga. But once Santa Anna was back in power, he wanted to fight the United States, and fight hard. This decision by Polk to return Santa Anna to power ultimately cost U.S. soldiers their lives at the Battle of Buena Vista. Though the United States technically won this battle, it came at the cost of 746 U.S. soldiers killed, wounded, or missing (pg 353, #4).

During this time, Mexico believed it would receive assistance for obtaining war products from Europe, which, in fact, did not happen (pg 21, #2). To compound Polk's problem, New England was firmly against the war because those states were very much against Texas becoming a slave state. From a military point of view, however, the U.S. army was doing very well in a hostile, enemy country. The United States continued to be victorious, at Palo Alto on May 8th, 1846 and in the Battle of Resaca de la Palma, with 1,200 Mexican fatalities versus 150 fatalities for America. This is amazing considering that the United States was outnumbered by the Mexican army 3 to 1. These victories came at the price of 34 fatalities at Resaca de la Palma and 113 wounded U.S. troops. A major factor in these American victories was superior firepower.

After these victories, in September 1846, the United States invaded Monterrey. It began at a place called Independence Hill and took place on September 22nd, 1846. Once that was successfully conquered and in American hands, the U.S. troops there literally had to fight house to house in Monterrey itself. Monterrey was the most heavily fortified city in Mexico. By September 25th, 1846, Monterrey was officially under military occupation, yet the U.S. army would lose 500 soldiers in this battle. As a result of this important military victory, Zachary Taylor was promoted to major general and became extremely popular at home, despite the fact that there were strong disagreements between Polk and Taylor.

Inspired by the military success of Taylor's forces, Ohio sent 3,000 troops to Texas, and Tennessee sent 30,000 men, in order to rush the Halls of Montezuma. Controversy arose when volunteers led by General Edmund Gaines arrived at Port Isabel, where they brutally raped, robbed, and violated the Mexican citizens (pg 542, #4). Mexican priests would call these American soldiers "vandal vomits from hell" (pg 543, #4). These savage, evil forces that represented America killed innocent Mexicans for fun, including children. The worst offenders of them all were the Texas Rangers, who were described as "packs of human bloodhounds." The Texas Rangers commander was Colonel Jack Hayes. The Rangers were known for being good fighters, but did things in their own fashion (pg 57, #6). Their enemy was Antonio Canales. They shot and killed 36 Mexicans in a ranchero near Agua Fria, and committed other acts of murder and carnage against the Mexican people, yet they were never punished.

By the end of 1846, almost all of northeast Mexico was officially in American hands (Matamoros, Reynosa, Camargo, Monterrey, Saltillo, Tampico, Tamaulipas, Nuevo León, and Coahuila) (pg 44, #2). The decision was made by President Polk to move against Mexico City by way of Vera Cruz (pg 45, #2). To lead this expedition, Polk had reluctantly named General Winfield Scott, due to Scott being a Whig, like Taylor. Yet Polk was determined that Scott would be the most qualified commander. This act by President Polk really upset General Taylor.

The Battle of Buena Vista took place on February 23rd, 1847, which the U.S. troops won after some fierce combat against their opponents. Buena Vista was a fierce battle, with 673 U.S. servicemen killed or wounded. During this battle, the U.S. troops observed that the land of New Mexico was poorly defended, and was controlled by corrupt Mexican officials. This inspired American officials to think that this land was available for acquisition. Once the United States decided to pursue this policy, the expedition to claim this land was given to Colonel Stephen W. Kearny and his 1,500 frontiersmen, who were known as "the Army of the West" (pg 57, #2). They advanced along the Santa Fe Trail, where Colonel Sterling Price led the expedition to Santa Fe and Kearny to California. In Santa Fe, there were massive revolts by the Mexican locals, including the assassinations of Governor Bent and several other high-ranking officials (pg 63, #2).

In response to these vicious attacks, Colonel Price organized a military unit whose primary intent was to crush the Mexican rebellion. This is precisely what it did. This U.S. military act silenced the uprisings in New Mexico for the rest of the war. Yet history shows that the acquisition of California was a much higher priority to the U.S. government. Commander John Frémont was the military leader given the assignment of claiming California (pg 64, #2), and he received the cooperation of the U.S. navy, which had ships stationed off the coast of California in case they were needed to assist the army invasion.

By July 7th, 1847, San Francisco had been successfully captured by U.S. forces. Commodore Robert Stockton followed this success with the successful capture of Los Angeles on August 13th, 1847 (pg 66, #2). Unfortunately, Stockton made the mistake of leaving a small American garrison stationed in Los Angeles, which resulted in a severe Mexican revolt. This riot resulted in the United States losing control of most of California. To compound the problem, Commodore Stockton was stuck in San Diego and was unable to assist in the recapture of Los Angeles (pg 67, #2).

To reclaim this territory, Kearny's troops came down to fight the Battle of San Pasqual, in which one third of his troops were killed (pg 68, #2). Despite this enormous loss, the United States still won this battle, and, eventually , in January 1847, Los Angeles was recaptured by America. To confirm this, Mexico and America signed the Treaty of Cahuenga.

Even though all of northern Mexico had been successfully captured by the United States, the Mexican government and army would still not give up. As a result, the U.S. army had to go to Mexico City to win the war (pg 71, #2). In order to achieve this, the U.S. army decided to sail troops to the Gulf of Mexico port of Vera Cruz to attack Mexico City (pg 72, #2).

Controversially, President Polk had passed giving command to General Taylor in favor of General Scott. Even though President Polk and General Scott were at odds politically, Scott had successfully landed 10,000 U.S. soldiers at the Vera Cruz beaches on March 9th, 1847 (pg 74, #2). Next, General Scott cut off incoming supplies to the city, while bombarding it with U.S. weaponry. By March 29th, 1847, Vera Cruz had surrendered. After this surrender, U.S. forces declared martial law there and provided rations to the inhabitants. The United States had fewer than 100 casualties in this battle.

Santa Anna waited in Mexico City for the U.S. military's arrival (pg 77, #2). When the two forces met, this became the Battle of Cerro Gordo, which the United States won by April 1847 (pg 78, #2). U.S. forces next successfully captured the city of Puebla and its 800,000 residents. Because of this, Santa Anna was having problems keeping his army together, and was losing support with much of the Mexican population (pg 83, #2). Although Santa Anna did consider resigning then, he instead became the absolute dictator of Mexico (pg 84, #2).

From the U.S. military's viewpoint, Mexico City itself could only be invaded a couple of different ways, because the city was surrounded by water. By August 20th, 1847, the invading U.S. army was located 3 miles from Mexico City. There was then a last futile effort to negotiate for peace, but this ultimately failed. Consequently, the Battle of Molino del Rey was fought on September 8th, 1847, which was one of the bloodiest battles of the entire war (pg 94, #2). This would cost over 700 casualties to U.S. forces and over 2,000 for Mexican troops for not a lot of territorial gain, but it did give the United States control of the Mexican city of Molino.

By September 14th, 1847, Mexico City also fell to U.S. capture and occupation (pg 98, #2 and pg 40, #3). Santa Anna had successfully escaped the city limits upon Mexico City's capture and consequently fled to Guadalupe Hidalgo. The U.S. military declared martial law in Mexico City in response to the large amount of rioting by Mexico City residents (pg 99, #2). This city's capture would finally be enough to break the will of the Mexican army to fight the Americans, and it consequently surrendered. By October 15th, 1847, Santa Anna had resigned (pg 100, #2).

Santa Anna fled to Jamaica but would later return to Mexico several times in the 1860s and 1870s. Both Generals Scott and Taylor had fought the Mexicans superbly, but both disagreed sharply with their civilian commander, President Polk. Polk had been forced to give the Vera Cruz/Mexico City campaign to General Scott, because he absolutely would not give it to General Taylor due to their political differences. Polk had called Taylor "a vindictive and ignorant politician" (pg 113, #2). Polk and Taylor were complete political opposites. Taylor found fault in Polk's military policies, and consequently thought that Polk was deliberately holding him back for personal reasons.

Polk did disagree with Taylor's decision to launch the Battle of Buena Vista, when he disobeyed the direct orders of the president. Polk vindictively did not congratulate him upon his victory. Instead, Polk credited Taylor's troops. Nevertheless, Taylor's popularity soared with the American public. Otis Singletary says in his book, *The Mexican War*, that "the Mexican War is a case history of poor civil-military relations" due to the poor relations that existed between President Polk and Generals Taylor and Scott. Singletary also says that the Mexican War involved two enemies: (1) Mexico and (2) ourselves (pg 128, #2).

Despite this, the results of the Mexican War were tremendously profitable for Polk's vision of "Manifest Destiny" for America. On February 2nd, 1848, the Treaty of Guadalupe Hidalgo was signed, effectively ending the war between Mexico and the United States and re-establishing peace between the two nations (pg 160, #2). Upon Mexico's surrender, it gave up 1,193,061 square miles of land, which is approximately five times the size of the country of France. This surrendered land acquired by the United States would become the states of Texas, Arizona, Nevada, California, Utah, New Mexico, and Wyoming, for which the U.S. government paid Mexico $15 million. This was done because Polk did not want to make it appear that he was conquering Mexico, but simply purchasing land. The U.S. Senate ratified the treaty on March 10th, 1848, by a vote of 38 to 14 (pg 161, #2). The Mexican War only lasted 16 months. By July 1848, the U.S. troop presence was gone. The war with Mexico was over.

The total U.S. casualties in this conflict were 13,780 killed, followed by thousands more wounded in action. While Polk kept his word and did not run for a second term, his political rival, General Zachary Taylor, was successful in capturing the presidency, which he claimed in the November 1848 election. Though Polk was sometimes criticized for being dull and stiff as the chief executive, he is also remembered as being a great achiever (pg 156, #3). Ironically, Polk died three weeks after he left office. Just as ironic, President Taylor died after 16 months in office from a case of cholera morbus, or severe cramps from poor sanitation. His replacement was Millard Fillmore.

One Congressman from Illinois, who served in the Black Hawk War, criticized his party's rival, President Polk, for the decision to lead our country to war with Mexico, which the Congressman deemed unnecessary. He claimed that the people had the right to choose their own form of government and should not be interfered with by the U.S. government. His name was Abraham Lincoln.

Bibliography

1. *From Sea to Shining Sea*, by Robert Leckie.
2. *The Mexican War*, by Otis A. Singletary (University of Chicago Press, 1960), disc 6.
3. *James K. Polk*, by John Seigenthaler (Times Books, 2003), disc 6.
4. *The Wars of America*, by Robert Leckie.
5. *The Complete Book of U.S. Presidents*, by William A. Degregario.
6. *Zachary Taylor*, by John S.D. Eisenhower (Times Books, 2008).

Chapter 7
The Civil War (1861-65) and President Abraham Lincoln's Actions and Decisions During These Years of Crisis

In all of the episodes of warfare that I have researched for this project, perhaps the darkest and most morbid years of American history occurred between 1861 and 1865. These years were dominated by the American Civil War, which was the bloodiest and most deadly war campaign in American history. Of the 1,556,000 Union soldiers who served valiantly for America during this time, 359,528 soldiers perished in combat, with an additional 275,175 men wounded. Of the Confederate soldiers who participated in this war, 258,000 perished, and an additional 225,000 men were wounded, out of the 850,000 soldiers who participated (#9). These grim statistics show that 41% of Union soldiers who participated to save the United States in the Civil War were killed or wounded, while 57% of Confederate soldiers suffered the same fate. The Civil War involved the most egregious battles ever fought on American soil, including Fredericksburg, Bull Run, Antietam, Gettysburg, Vicksburg, and Cold Harbor.

The war itself was fought over the extremely controversial issue of slavery, and slave rights for the southern states. At the time, the southern economy was heavily dependent on having slaves work on the plantation, whose goods produced by slaves were then sold to other nations, as well as other parts of the country. The institution of slavery in America began in August 1619, when Virginia tobacco farmers purchased 20 slaves from Dutch traders. Also in 1619, the Spanish and Portuguese slave owners had transported more than one million Africans to the Caribbean and South American colonies (#9).

The United States soon started importing huge numbers of African slaves, who were wrongfully imprisoned in Africa and shipped in inhuman conditions to the white plantation owners. Most northern states did not agree with the concept and practice of slavery. In fact, Rhode Island first abolished slavery in 1774 (#9). This unethical practice of slavery caused controversy and animosity between northern and southern states. The southern economy, which depended heavily on the plantation industry, was completely engulfed in the practice of slavery in order to financially succeed in developing crops and agriculture. These products garnered heavy use in America and other foreign markets. In fact, the invention of the cotton gin by Eli Whitney revolutionized the cotton industry by giving the southern plantation and slave owners something economically profitable to work with. After all, two of the founding fathers of America, George Washington and Thomas Jefferson, had slaves in their respective homes (Mt. Vernon and Monticello).

More controversy erupted between slave owners and anti-slave citizens when a protective tariff law was passed in 1828 that heavily taxed manufactured goods from abroad. This law was intended to discourage importation/exportation of international products. Therefore, the intention was to encourage more sales of domestic goods within the United States. This law would hurt the south economically. Yet when Andrew Jackson won the presidency in 1828, he promised to bring tax reform with him to the White House. Jackson was known for his love of the United States, and he wanted the states to remain together, no matter what. Yet, unfortunately for the southern states, Jackson's tax reform laws of 1832 proved to be unhelpful.

The southern states grew frustrated with the federal government's apparent unwillingness to assist (#9). The southern representatives in Congress started to debate ideas and concoct strategy, such as nullifying federal laws in their states, or even outright secession from the Union (#9). However, leaders like John Calhoun saw that someday slavery would be abolished by a northern Congress. So Calhoun and his representatives wanted slavery protected at all costs. Unfortunately for Calhoun, who was Jackson's vice president during his first administration, he was informed by the president that federal law takes precedence over state law, and Jackson would use the military to enforce the laws implementing the tariffs if he had to. A statement like this from President Jackson was no idle threat.

Despite the practice of slavery in the beginning of the 19th century, emancipation of all slaves was gaining momentum nationally, and William Lloyd Garrison had predicted the end of slavery as early as 1831 (#9). A powerful anti-slavery crusader named Frederick Douglass gained popularity as an escaped slave, and he told the world of the horrible, awful cruelties of the slave industry. Douglass became the new African-American voice of anti-slavery (#9).

By the year 1861, of the 12,000,000 people in the southern U.S. population, one third were slaves (#9). In an effort to combat the slave trade, the underground railroad, created by white abolitionists, was successful in smuggling some number of slaves to freedom. However, those who were caught, both black and white, were severely punished (#9). President James Buchanan, during his term (1857-61), was criticized for not having the moral strength and leadership to prevent the final occurrences that set the stage for the American Civil War.

During the election of 1860, lawyer, former congressman, and former Black Hawk War veteran Abraham Lincoln won the Republican nomination and went on to face Democrat Stephen Douglas for the office of President of the United States. Douglas was in the same war as Lincoln, and narrowly defeated him in the 1856 Senate race in Illinois. Even then, Douglas supported slave rights for states, while Lincoln was known for being a strong anti-slavery candidate. In the presidential election of 1860, Lincoln prevailed over Douglas, with 1,866,452 votes compared to Douglas's 1,376,957 votes. Though Lincoln could enjoy the moment of victory after a long climb and briefly bask in his many accomplishments that propelled him to the pinnacle of power, he hardly had a peaceful unification process from his country to rally him into a fresh, new administration.

To the contrary, once Lincoln was elected president, and before he was sworn in as commander-in-chief, seven states seceded from the union, starting with South Carolina on December 20th, 1860 and followed by Mississippi on January 9th, 1861, Florida on January 10th, 1861, Alabama on January 11th, 1861, Georgia on January 19th, 1861, Louisiana on January 26th, 1861, and Texas on February 1st, 1861. With this monumental challenge that Lincoln faced at the beginning of his term, and the bloodiest war ever fought on American soil that would ensue for his next four years, it is important to understand Lincoln's life and career before the presidency, and the experiences that led him to the nation's highest office. I will also examine the decisions that Lincoln made as president that were militarily successful in the Civil War, which literally threatened to tear the nation into two separate entities. This war would cost over 600,000 lives on both sides of battle, and it would also cost Lincoln his own life as soon as the war had ended—from a Confederate sympathizer assassin's bullet in Ford's Theatre in Washington, D.C. But in order to fully comprehend the American triumph and tragedy that occurred during this period, I will now examine Lincoln the person, who would eventually become one of our country's finest leaders.

Abraham Lincoln was born on February 12th, 1809 in Hodgenville, Hardin County, Kentucky. He was named Abraham after his grandfather, by his father, Thomas, who lost his father from an Indian attack in Kentucky in 1786. When Abraham was seven years old, his family moved from Kentucky to Indiana, because Kentucky was admitted to the union as a slave state. Slavery was a practice that the Lincoln family did not believe in. Abraham, thanks to his family, learned about the incorrect , immoral behavior of the slave trade at an early age. Living on a farm led to somewhat of an Amish lifestyle, and Lincoln worked tirelessly for his family by providing planting and field work on it. In fact, Lincoln sustained a serious injury in an accident on that farm as a youth, when he was kicked in the head by a horse, which nearly killed him (pg 26, #4).

Lincoln learned to tell stories well by listening to his dad tell good stories to crowds of people when Abraham was a boy (pg 51, #12). Later, when Abraham was nine years old, things went from bad to worse when his mother, Nancy, died suddenly on October 5th, 1818 after a probable case of brucellosis. While Abraham was in deep mourning over the loss of his mother, his father Thomas courted and married Sarah Bush within a year of Nancy's passing. Sarah had three small children of her own, and Sarah treated both her children as well as Abraham and his siblings as equals. This helped Abraham through the grief of losing his mother.

As far as Abraham's formal education is documented, it is reported to have lasted until about the age of 15. During his education period, Lincoln attended Crawford School for one term, which is three months. Then Lincoln followed Crawford by attending a school run by Azel Dorsey in the same cabin that the Crawford School had used. Though Lincoln's educational years seem to be remarkably short and simple by today's standards, his education allowed Lincoln to master what he called "reading, writin, and cipherin' (pg 29, #4). These skills that Lincoln absorbed would come to pay off handsomely in his quest for the presidency.

Yet, at the time, Lincoln's career path seemed to be in the farming industry. Lincoln despised the farming trade, and he wanted that world eradicated from his life by his 21st birthday. Lincoln's main interest was in public speaking, something Abraham's father, Thomas, would never have done. Lincoln had an extensive employment record during his twenties, which included work as a carpenter, riverboat man, store clerk, soldier, merchant, postmaster, blacksmith, surveyor, then lawyer, and finally politician. Lincoln served in the U.S. Army during the Black Hawk War.

After serving his country, Lincoln ran for state legislator in Illinois during 1832. He competed for the position against 12 other candidates, the top four of whom were elected. Lincoln finished eighth and did not win a seat . However, two years later, in 1834, Lincoln won the same election he lost in 1832 by finishing second. Lincoln was thoroughly consumed with reading books, and he literally read everything that he could get his hands on. Lincoln's interest in practicing law began to really develop and flourish during this time. Lincoln also loved to debate. In fact, Lincoln began his historic series of debates against Stephen Douglas as early as 1838, and he later debated Douglas for the senatorial and presidential offices in 1856 and 1860, respectively. These two men would debate over issues such as national and state banks, which Lincoln argued for and Douglas against. These debates over important issues for his time strengthened Lincoln's grasp of Republican issues and platforms, namely, anti-slavery and union.

Lincoln also spoke out against mob violence when, in 1837, a man named Elijah Lovejoy was murdered by an anti-abolition crowd in the state of Missouri. While Lincoln did not condemn the attackers by name, he did condemn the people who "throw printing presses in the river, and shoot editors." Lincoln made this comment because Lovejoy was an editor of the *Observer*, which was an abolitionist newspaper.

Lincoln became one of the most prominent lawyers in the state of Illinois during his law-practice years. It was during this time that he courted and married his wife, Mary Todd. Lincoln won his election to the U.S. House of Representatives from the state of Illinois in 1846. His strongest beliefs for government during this time were for a national bank and federal money for interstate improvements.

During Lincoln's time in Congress, he did openly criticize President Polk and the war with Mexico. He also questioned America's role as a federal power once it was over. Lincoln's experience as a member of Congress exemplified his skills from the legal industry. When Lincoln's term in Congress expired in 1849, he returned to private law practice in Illinois. Lincoln's skills as a lawyer began to pay off quite lucratively for his family. He took on a heavy load of court cases in order to bring in the substantial money that he and Mary desired. Lincoln earned the nickname "Honest Abe" for his court reputation of being absolutely honest.

Lincoln's price for his fame in the legal world was profound, and there was severe tension between Lincoln's law partner, William Herndon, and Lincoln's wife, Mary. The problems of Mary not getting along with Abraham's associates persisted throughout Lincoln's personal and professional life, and caused him great strife. Ironically, even Mary's own family members were Confederate sympathizers during the Civil War.

While Lincoln enjoyed a return to his private law practice, he missed serving the people of Illinois in Congress. He felt the urge and desire to return to politics. In 1852, Lincoln served as Whig National Committee chairman. That year, the Democratic nominee was Franklin Pierce, who had close ties to Stephen Douglas. Lincoln's own anti-slavery views were shared by his friends in Congress, such as Joshua Biddings and Horace Mann (pg 165, #4). Due to the common belief at the time that blacks and whites could never live together peacefully as a nation, Lincoln's initial policy on slavery was to ship all African-Americans to a place called Liberia, so that they would live freely there (pg 166, #4).

Yet this notion proved to be impractical, for the slaves were not going to be set free and did not want to be migrated forcefully to another continent. When the Kansas-Nebraska Act was enacted, which opened these two states to slavery, it was based on the supposed "popular will" on that issue of the citizens of these two states. This caused a huge uproar of protest, with men such as Salmon Chase and Cassius Clay strongly denouncing adding slave states to the west. Surprisingly, Lincoln stayed silent in the dogmatic political fight over slavery that was occurring in Washington, D.C., but he did not hold office at that time.

In 1853, Lincoln allied himself openly with the abolitionist cause in order to repeal the Missouri Compromise, which had already been signed into law. Lincoln once said, "If slavery is not wrong, then nothing is wrong" (pg 91, #12). Lincoln's political opponents, led by Stephen Douglas, argued that the law was "superseded" by the Compromise of 1850, which had the authority to give popular sovereignty to the territories, called "self-government" (pg 173, #4). Though Lincoln was fiercely anti-slave in private, he surprisingly did not take a stronger alignment with the Radicals, who called for an immediate end to all slavery in all territories. Lincoln's feeling during this time was that he was willing to accept some form of slavery in certain states because he "would consent to the expansion of it rather than see the Union dissolved" (pg 201, #4).

It could be suggested that Lincoln had affection for the south, considering he was from the state of Kentucky. As mentioned before, Mary Lincoln had pro-slavery family members. Though Lincoln disagreed with the Dred Scott decision that was administered by the Supreme Court, he reacted too slowly to make an impact for the anti-slavery movement. The Dred Scott case by the Supreme Court found that blacks were not citizens under the Constitution, and therefore did not have the same rights as other Americans (pg 189 , #12). Despite this, Lincoln openly disagreed with Supreme Courts Chief Justice Roger Taney's opinion that the Constitution and Declaration of Independence did not apply to black people, because Lincoln rightfully believed that they did. This caused Stephen Douglas to go on a verbal-assault campaign against Lincoln, and he accused him of wanting to "eat, sleep, and marry with Negroes" (pg 202, #4).

Consequently, Lincoln rejected this accusation by Douglas by stating that "all men are created equal, equal in certain inalienable rights, among which are life, liberty, and the pursuit of happiness" (pg 202, #4). Unfortunately, Democratic President James Buchanan accepted Kansas as a slave state because, out of the 9,000 eligible voters in the state, only 2,200 actually showed up to vote. As a result, slavery was passed and Buchanan signed the bill into law. As early as 1855, Lincoln raised the question, "Can we exist as a nation half-slave or half-free?"

One of the most infamous campaigns for a Senate seat in American history occurred in 1856, when candidates Abraham Lincoln and Stephen Douglas met in a fierce competition to determine the winner—and the future of Illinois's vote on the slavery issue in the U.S. Senate. Douglas was a Democrat with pro-slavery views, and Republican Lincoln represented the Republicans, who held anti-slavery positions. In order to wrench this Senate seat from his formidable opponent in 1856, Lincoln had to make the Illinois campaign versus Douglas about three central issues:

1. National slavery had to be defeated
2. Stephen Douglas powerfully contributed to that institution
3. Therefore, Douglas had to be defeated!

Contrary to Lincoln's platform, Douglas and his constituency supported slave rights and those who agreed with this horrible degradation of man. It was upon this platform that Douglas hoped to rally votes.

Lincoln and Douglas agreed to seven debates: Ottawa, Freeport, Jonesboro, Charleston, Galesburg-Knox College, Quincy, and Alton. These historic debates included discussions about issues such as slavery and the Freeport Doctrine. The Freeport Doctrine proposed that people of a territory can, by lawful means, exclude slavery from their limits prior to the formation of a state constitution (pg 218, #4). Both Lincoln and Douglas debated each other on these issues vigorously. Unfortunately for Lincoln, though the Republican party could claim victory in the popular vote in the state of Illinois during the election of 1856, during which its citizens could elect their candidates for state treasurer and superintendent of education, it could not gain control of the state legislator who controlled the vote for the Senate. Of the 13 votes that were able to determine the electors, 8 of them belonged to the Democrats, who elected Douglas over Lincoln. Though Lincoln was bitterly disappointed by his defeat, and thought that his political career might be in ruins, he vowed to fight on for civil rights for all Americans, of all races (pg 229, #4).

Contrary to Lincoln's view, Senator Douglas vowed to undermine resistance to the spread of slavery, so that excluding slavery by state laws would soon be overruled. This national slave code could be enacted. The issue became slave rights for citizens of Illinois. Senator Douglas's Democratic message was that African-Americans were not covered by the Constitution, and that slavery was an institution that would be protected by Democrats. Douglas was a popular Democratic member of the Senate, and the frontrunner for the Democratic nomination for president in 1860. The Republicans knew that they needed to nominate someone who was a westerner, so that they might have a chance to carry the west, where Douglas was hugely popular. Lincoln fit that profile well, and he was nicknamed the "Rail Splitter" for his father Thomas' involvement in pioneering Macon County. This is similar to President Jackson's nickname, "Old Hickory," or President Harrison's nickname, "Tippecanoe."

The Republican Party then was a mixture of Whigs, Know-Nothings, and some anti-slavery Democrats (pg 239, #12). When the Republican convention met in Chicago in 1860, they discussed issues such as endorsing the Homestead Act in an effort to help farmers in the west.

Federal help was also needed to improve rivers and harbors, and to improve cities of the Great Lake region, such as Detroit and Chicago, Tariffs were also needed to protect the iron interests of Pennsylvania and New Jersey (pg 246, #4). When the time came to select a nominee, Republicans were looking for a candidate who would be strong in states such as Pennsylvania, Indiana, and Illinois. Lincoln stood out for his strong appeal in Illinois, and he was known as a strong anti-slavery candidate. Lincoln's popularity won him the Republican nomination on the third ballot in 1860, with William Seward as his closest competition (pg 252, #4).

Soon the political attacks would surface, including Lincoln's previous opposition to the Mexican War by way of failing to vote for supplies to the American troops in the field. Lincoln would also be accused of slandering the memory of Thomas Jefferson by accusing Jefferson of "pulling out of liberty, equality, and the degrading cause of slavery" (pg 253, #4). In the meantime, the Democrats were split on their nomination of Stephen Douglas. Although Douglas had strong support in the north, a man named John Breckinridge was winning support in the south.

In the end, Douglas beat out Breckinridge for the nomination. That meant that Lincoln and Douglas would rekindle their famous debates that left off in 1856, and they carried on with velocity in 1860. On election night, Abraham Lincoln defeated Douglas by receiving 1,866,452 of the popular vote, compared to Douglas' 1,376,957. But unfortunately for Lincoln, he assumed office at a time when the country was disintegrating. In fact, President Buchanan, on his way out of office, seemed resigned to accept the secession of the southern states, and others were ready to accept their peaceful departure from the union. Not Abe Lincoln!

As Lincoln assumed the presidency, his rival Stephen Douglas met with him and pledged his full support for Lincoln to preserve the Union (pg 348, #12). At the time, Lincoln was the youngest man elected President of the United States. He was 51. This fact seemed to some people a political liability. As the country was literally splitting in half, Lincoln was forced to take immediate action. Lincoln was not, in any way, amenable to the secession of any state from the Union, and he strongly believed that it must be preserved. Some northerners saw secession of the southern states as "terrorism" by the "traitors of the south" (pg 261, #4).

Lincoln selected all of his one-time political opponents, such as William Seward, Edward Bates, and Samuel Chase, as Cabinet members. General Edwin Stanton was named secretary of war. Meanwhile, Jefferson Davis was chosen as President of the Confederacy for a supposed six-year term. Ironically, Jefferson Davis was once President Franklin Pierce's secretary of war. Alexander Stephens of Georgia was selected Vice President of the Confederacy. The states that still had slaves but did not secede included Delaware, Maryland, Kentucky, and Missouri

(#9). In fact, West Virginia was carved out of Virginia because it remained loyal to the Union, and thus became a state on June 20th, 1863 (#9).

Lincoln's vice president was Hannibal Hamlin of Maine. Upon entering office on March 4th, 1861, the situation had gone from bad to worse regarding the Union's stability. By the end of April 1861, the states of South Carolina, Florida, Mississippi, Texas, Louisiana, Georgia, Virginia, and Alabama had seceded from the Union and joined the Confederacy. While Lincoln was willing to accept some form of slavery in the Union where it had previously existed, he would absolutely not accept any states leaving the Union (#9). At first, Lincoln and his Cabinet had declared war against the seceding southern states—not to end the practice of slavery, but to preserve the Union (pg 370, #12). Lincoln believed that there were two ways to settle a domestic grievance that would change the national government. One way was to amend the Constitution, and the other was through revolution (pg 268, #2).

Lincoln adopted the philosophy that the Union must be saved at all costs. Yet even on his way to his own inaugural address, Lincoln felt it necessary to speed through Baltimore, Maryland, due to the threats on his life from the pro-slavery state. Nevertheless, Lincoln was inaugurated on March 4th, 1861, protected by heavy security. As he walked to the Capitol building with President Buchanan, Lincoln admitted that "he was entirely ignorant not only of the duties, but of the manner of doing business as the President."

The Civil War itself began on April 14th, 1861, at the Battle of Ft. Sumter. This battle was considered a Confederate victory. Yet Lincoln, though he was new to the presidency, had to feel good about the superior numbers the Union and its army had over the Confederates. At the time, the Union had approximately 20 million residents, as opposed to the 5 million citizens of the south. Some members of Lincoln's administration, including Seward, felt that the war would be over in 90 days. Accordingly, those who enlisted for the Union army signed up for 90 days of service.

The Union could not have been more wrong in its ill-advised prediction as to the Confederates' will and determination. On top of the loss at Ft. Sumter for the Union, more problems surfaced when the state of Maryland sided with the Confederacy, which made northern travel to Washington, D.C. more difficult through Maryland (pg 298, #4). Though this seemed to help the Confederate cause, Lincoln countered with the suspension of habeas corpus. Lincoln had the power to enforce this rule under Article 1, Section 9 of the Constitution. This meant that Union soldiers could summarily arrest people merely helping, or even attempting to help, the Confederacy. Such a person, under the suspension of habeas corpus, could be held in custody indefinitely (pg 299, #4).

The next battle was the first Battle of Bull Run, which was a Confederate victory that took place in Centerville, Virginia, approximately 18 to 20 miles southwest of Washington, D.C. (#9). Nine hundred Union soldiers would perish in this battle (pg 371, #12). Confederate General Pierre Gustave Toutant-Beauregard was the victorious commander in this battle, but both sides suffered massive casualties. It was the battle that spawned the story of Thomas "Stonewall" Jackson. The defeated northern General Irvin McDowell was replaced by General George McClellan, otherwise called by the press as "young Napoleon" (#9).

McClellan had graduated second in his class at West Point in 1846, and he fought bravely and successfully for the United States during the Mexican War with General Winfield Scott. When McClellan was first appointed to command by President Lincoln, he successfully won the Battle of Philippi on June 3rd, 1861. This was followed by another victory for McClellan at the Battle of Rich Mountain on July 11th, 1861. Because of this, Lincoln gave McClellan the command of the Army of the Potomac (#9).

Despite this, at the beginning of the war versus the Confederacy, the Union army was poorly prepared and poorly led. This fact was heavily amplified by the Union loss at the First Battle of Bull Run. A decisive Union victory there could have theoretically ended the war.

While the Civil War was going on, the House still included southern leaders, and the states in secession were never recognized as independent states that belonged to another country. In other words, the House, executive branch, and federal government tried to carry on as normally as it could, despite the most extraordinary challenges and upheaval in the country's history.

Ironically, Stephen Douglas, Lincoln's Democratic opponent in the 1860 presidential election, died on June 3rd, 1861. This fact might have drastically altered America's future had he, and not Abe Lincoln, been elected to the presidency.

While he was in desperate need of support at home, Lincoln received the news that the "war Democrats" rallied behind him in his effort to save the Union. Union General John Frémont wanted to seize property and slaves from the state of Missouri in an act of martial law. Lincoln, however, overruled him. This would cause conflict between Frémont and Lincoln. General Frémont did not perform well militarily in the beginning of the war; therefore, Lincoln fired him. Lincoln did this because he needed Union army victories before he could start considering emancipation for the states. Lyman Trumbull also wanted to confiscate property and slaves, but Lincoln could not allow these men to set policy on slavery.

Problems for the Union were amplified with the Confederate army victory at Ball's Bluff on October 21st, 1861. The grim casualty results from this battle were 49 Union troops killed, 158 wounded, and 714 captured or missing. The Confederacy had 33 soldiers killed, 115 wounded, and one missing. Among the dead for the Union included Senator Edward D. Baker, who had volunteered for service (#9).

Attorney General Edward Bates recorded in his diary about President Lincoln, "The President is an excellent man, and in the main wise, but he lacks will and purpose, and I greatly fear he has not the power to command" (pg 328, #4). General McClellan would make matters worse when he would delay his commitment of attacking the Confederates, because he always overestimated his opponents' numbers of troops. This fact, combined with Generals McClellan and Henry Halleck's failure to win the war, led Lincoln to actually consider leading troops into battle himself, despite his lack of military training.

Meanwhile, back in Washington, a group of men calling themselves the Jacobins constantly attacked Lincoln on his policies. The group consisted of Senators Benjamin Wade, Zachariah Chandler, and Lyman Trumbull. In January 1862, the Union army obtained substantial victories in two strategically important places: Forts Henry and Donelson. General Ulysses Grant was credited with these victories. These were desperately needed Union army victories in the south.

As a result of General Grant's victories, Grant was given control of the Army of Indiana. Grant's name originally was Hiram Ulysses, but when he attended West Point, his name changed to Ulysses S. Grant (#9). Grant was born on April 27th, 1822, at Point Pleasant, Ohio. This was during President Monroe's "era of good feeling" (pg 2, #11). He had three sisters and two brothers. Despite Grant having attended West Point, he was not considered a scholar. However, Grant's victories at Forts Donelson and Henry brought Tennessee and Kentucky back to the Union.

This was the most significant surrender to U.S. forces since Yorktown (pg 164, #11). As a consequence, Republicans in the House wanted Lincoln to condemn slavery and ban it lawfully. For example, Trumbull put up a Confiscation Bill before the House, which would have eliminated slave practice forcefully, but Lincoln overruled him. He did this because although he condemned slavery as an institution, he was willing to put up with it if it would prevent some border states from revolting.

Lincoln considered the question, If freeing slaves was used as a military tool, where should the freed slaves be sent? One suggestion was relocating them to such places as Haiti, Central America, the Danish West Indies, the British Honduras, and Dutch Guiana. Yet when other foreign governments saw the struggle of the war being waged in America over freedom versus slavery, these countries sided with the Union. This would drastically increase the chances of the Confederacy being defeated.

During this time, however, General George McClellan was at serious odds with the Lincoln Administration. McClellan had 100,000 Union troops at his disposal, but he was reluctant to use his men to aggressively attack the enemy. This was because McClellan believed, incorrectly, that the Confederacy actually had more manpower than it really had. The Army of the Potomac, under General McClellan's leadership, had high morale but was acrimonious to the president. Lincoln believed that the Union army should have already defeated the Confederate army.

Lincoln's position on slavery radically shifted when a group of Quakers suggested to him, during a White House visit, that he should emancipate all of the slaves. Lincoln called slavery "the greatest wrong ever inflicted on any people" (pg 469, #12.). While Lincoln agreed with this suggestion, he felt that he did not have the proper authority to carry out such a proclamation while the southern states were in open revolt. Lincoln was similarly having trouble selling the Union cause to new recruits and northern citizens sympathetic to his struggles. This is largely blamed on McClellan's failure in battle, which caused the Union military to lose momentum. Lincoln did not have experience with military command and lacked the military technological comprehension to issue military orders. Lincoln was a politician, lawyer, and veteran of the Black Hawk War, but he was never a military commander. Sensing a weakness from the northern army, and intent on forcing the Union to negotiate for peace, General Robert E. Lee chose to invade Maryland in 1862 (#10). Karl Marx would comment, years after the conflict, that he thought Antietam was the event of the entire war.

The brutal and intense Battle of Antietam began on September 17th, 1862, in a place called Sharpsburg, Maryland. General Robert Lee saw that if he went on the offensive, he would catch the northern army off guard. This bold military initiative might be the only way for the south to win the war. At the time, Maryland was considered a neutral state. Lee ordered his army to cross the Chesapeake on September 4th, 1862. After all, the Confederacy had momentum from the Manassas victory going into Antietam. Lee had incentive to invade rich, northern farms in order to acquire badly needed supplies and food. On the global stage, European nations were waiting to see whether the south warranted their support.

To stop Lee's army's advance in Maryland, Lincoln once again turned to General McClellan. McClellan and his forces moved west to Antietam. Lee had set a trap for the Union to fall into. At that time, Lee's forces outnumbered McClellan's two to one. The sunken road between farms became known as "Bloody Lane." At 9:30 a.m., the 2nd Corps of the Army of the Potomac moved in with 3,000 more troops. Commander William French's troops led the division for the north. The Union troops then advanced on the Confederate army. Confederate General John Gordon waited to fire on advancing Union troops until the last second. When the firing did begin, 440 troops died in five minutes.

The 2nd and 3rd Brigade moved in next. The Confederate army began to weaken. General John Gordon himself was shot. French's brigade was badly beaten also. General Israel Richardson was called in to lead the Irish Brigade for the Union. In this battle, the Irish Brigade took extremely heavy casualties, with half of the army dying in the charge. The

Irish troops did finally get support from the Union army, but 540 members were killed in the process.

At 11:00 a.m. on September 17th, Union Commander John C. Caldwell and his troops had finally arrived. Now, the tide of battle turned to the Union, and the union army was now winning. Many Confederate soldiers were killed. General Gordon for the Confederates was shot five times, but he survived. Though General Lee wanted to strike the Union army while the iron was hot, he miscalculated in his belief that the Marylanders he encountered would support the Confederate cause. This may have been because most citizens of Maryland, who supported the south, did so at the beginning of the war.

Though General Lee brought 50,000 soldiers with him into Maryland, they were not enough to defeat the Union army. Yet the fatality/casualty rates on both sides were horrific; 12,000 men were wounded in one day. This included members of the Irish Brigade, who lost many soldiers on the horrible site of Bloody Lane. The Confederates lost between 1,900 and 2,700 men on the first night at Antietam. The Union army lost approximately 2,108 dead and 9,000 wounded. The difference seemed to be that McClellan received 20,000 reserve soldiers on the day of battle, whereas Lee received none. Yet Lee successfully eluded capture with his troops. Upon learning this information, Lincoln was very disappointed. A decisive, overwhelming military victory could have ended the war at that stage.

Despite this, Antietam did give the Union the great hope of eventual victory. This was even enough to give Lincoln the courage to release the slaves as a sign of divine inspiration from God. Lincoln did this in spite of his disappointment with the Union not finishing off the rebels, once and for all. Because of Lincoln's emancipation order, the British government never recognized the Confederates. In fact, England praised Lincoln for releasing the slaves. This also had big consequences in the 1862 political elections. The Republicans kept control of the House and a majority of governorships, as well as gaining five seats in the Senate. The Democrats had been predicted to win in a landslide before the Battle of Antietam took place.

Unfortunately, after the successful Union victory at Antietam, McClellan was foolish enough to do nothing for six weeks. Lincoln, justifiably angry, even visited McClellan on October 15th, 1862, and told him to find motivation in pursuing Lee. McClellan openly defied that order. By the time the Army of the Potomac did get going on October 26th, 1862, General Lee was long gone to protect the city of Richmond, Virginia. As a result of General McClellan's actions and inactions after the Battle of Antietam, Lincoln consequently fired him on November 7th, 1862.

General Ambrose Burnside took over command. One more punch from McClellan could have ended the war, but instead, Lee and "Stonewall" Jackson regrouped and launched new attacks. However, Antietam was called the turning point in the war for the Union. As a result, most southern citizens were becoming depressed and confused by Confederate leadership.

The Emancipation Proclamation that was issued by President Lincoln said, "As Commander-in-Chief of the Army and Navy of the United States, as a fit and necessary military measure, all persons held as slaves in any state, where in the Constitutional authority of the United States shall not then be practically recognized, be forever free." This act by Lincoln brought criticism from the Democrats. General Burnside, who replaced General McClellan, would do no better in achieving military victories. The Union army, however, did win the Battle of Corinth, Mississippi on October 3rd and 4th, 1862. This victory was secured by Generals Grant and William Rosecrans.

Meanwhile, General Burnside, with his Army of the Potomac troops, suffered a huge defeat when they crossed the Rappahannock River into Fredericksburg, Virginia. It was there that the Confederate army lay in waiting for the Union troops. John Pope would be responsible for the loss at the Second Battle of Bull Run, which occurred in 1862. Pope was beaten by a surprise counter-attack, and he lost five times as many men than were lost at the First Battle of Bull Run. After enduring losses from Frémont, McClellan, and Pope, Lincoln watched Burnside make the mistake of delaying attacking Lee's army at Fredericksburg, and Lee's troops were waiting in trenches in order to ambush the Union troops. The battle for Fredericksburg was important for both sides, because it was located halfway between Washington, D.C. and Richmond.

The battle took place on December 13th, 1862, and it was a total massacre for the Union troops at the hands of the Confederates. General Burnside would continue to send his men wrongly into battle, where they were slaughtered by Confederate gunfire. Of the 106,000 Union troops who participated in the Battle of Fredericksburg, 12,700 were killed or wounded. Of the Confederate troops who participated (72,500 men), there were 5,300 killed or wounded (#9). Fredericksburg was the worst defeat in the U.S. army's history (#9).

While the carnage of the Civil War was going on between the north and the south, the Union also had to deal with major uprisings and combat from the Native Americans in the territory of Minnesota. There were also Native American uprisings in the west, in places like Arizona and New Mexico. Union forces quelled the uprisings in the west, but not without some controversy. The "Sand Creek Massacre" occurred in Colorado in November 1864, when Union troops gathered a large group of natives and killed them all, including women and children (#9). This act of horror may have been in response to the earlier uprisings in Minnesota that occurred in 1862.

It was the Santee Sioux tribe that attacked and killed many white settlers, due to the controversial Mendota Treaty of 1851. That treaty stipulated that white settlers would provide to the natives all flour and beef products, in exchange for white colonization of the land. But instead of flour and beef products, the natives received pawned-off spoiled pork and flour at expensive prices, while the white settlers took control of the Sioux's best hunting grounds and wheat fields. The Lincoln Administration seemed unaware of the atrocities in Minnesota, until Lincoln sent his secretary, John Nicolay, to Minnesota in July 1862 to investigate the matter.

Since no native attacks had taken place as of Nicolay's visit in July, there was no recommendation to the president for troop presence in Minnesota to protect the settlers, due to their urgent need to be in the east to fight the rebels in the Civil War. The white settlers felt that they had a trustworthy liaison to deal with named Little Crow, who was considered a friend. But the attacks by the tribes began when Little Crow assembled the "Seven Council Fires" leaders, which were the Wahpeton, Wahpekute, Sisseton, Yankton, Yanktonai, Teton Sioux, and Santee Sioux, to unite with his tribe and declare all-out war on the whites.

It began on August 17th, 1862, with an attack at Acton Post, with natives killing five whites, including a child. The next day, Monday, August 18th, the natives launched a vicious attack against the whites, killing 200 white farmers and boys in the first 24 hours and capturing all females. Another attack occurred the following day at Lake Shetek in southwest Minnesota, which wiped out half of the whites. The Sioux then attacked Fort Ridgely, but were repulsed by cannon fire. The attacks continued into Wisconsin, Iowa, and the Lake Superior region.

Governor Alexander Ramsey of Minnesota informed Lincoln that half of the state's population were fugitives. Lincoln, at that point, already had his hands full with the disaster of the Second Battle of Bull Run. Secretary of War Stanton informed Lincoln that he would release paroled Union prisoners from St. Louis to go to Minnesota. Colonel Henry Sibley took charge of 1,400 troops who knew the language and culture of the Sioux well. His troops defeated the Sioux in the Battle of Yellow Medicine River and the Battle of Wood Lake. The U.S. soldiers recaptured land held by the natives and released all prisoners held there. Four hundred Sioux males were arrested and 392 were legally tried there.

Of the 392 who were tried in court, 306 were sentenced to death by hanging. Lincoln asked for a personal review of the 306 death sentences, and reduced the number of these condemned natives to 39. He did this despite the fact that his own grandfather was killed by Native Americans in Kentucky in 1784. Of the native prisoners whom Lincoln spared from death, he ordered them protected from any "unlawful violence" (pg 124, #2). This act by Lincoln outraged many whites, who thought that the natives did not deserve such sympathy. Yet 39 native prisoners were hanged.

The Sioux were defeated at the end of 1862, and Congress signed a treaty with the tribes. Little Crow himself was killed by a settler in 1863. By 1863, "peace had come to stay in Minnesota" (pg 126, #2). By January 20th, 1863, Lincoln had replaced General Burnside after the Fredericksburg disaster with "Fighting Joe" Hooker (#9). Hooker was from Massachusetts, and he graduated from West Point in 1837. Hooker had previously served in both the Second Seminole War (1835-43) and the Mexican War. He fought alongside Generals Lee and Jackson in these campaigns. Hooker improved the conditions of the Union army in the beginning of 1863.

The Confederates were having problems of their own, suffering many military desertions, including an estimated 50,000 to 100,000 Confederate soldiers after losses at Gettysburg and Vicksburg in 1863. Also, Confederate money was practically worthless. This fact made economics very hard for the south. Union forces did suffer defeats in the beginning of 1863, at the Battle of Charlestown and in eastern Tennessee. Meanwhile, in the south, Union Generals Rosecrans and Grant defeated Confederate troops in Corinth, Mississippi. Yet again, the Union failed to finish off the fleeing Confederate troops. Grant saw the importance of capturing Vicksburg, which would isolate the Confederacy east and west of the Mississippi River.

Yet Vicksburg was a heavily fortified post, with weapons from the Confederates to prevent such an attack (#9). In fact, in December 1862, General Grant led a failed attempt to capture Vicksburg (pg 225, #11). Yet Grant would show Lincoln that unlike McClellan, he had the will to aggressively pursue the enemy, as he proved at the Battle of Shiloh. When Grant was beaten on the first day of Shiloh, which included 1,000 Union troops killed, wounded, or captured, Grant regrouped and reclaimed the territory the next day. This battle would come to ultimately doom the Confederate cause in the Mississippi Valley.

Grant won a bloody battle at Jackson, Mississippi, called Champion Hill, after heavy losses on both sides (#9). This would help set up the Union attack on Vicksburg after a second failed attack on the city. Vicksburg finally fell to the Union after heavy bombing of the entire city with Union artillery. The Union, once Vicksburg was captured on July 4th, 1863, had full control of the Mississippi River (#9). This would, ironically enough, happen the day after the Battle of Gettysburg, which occurred in the fields of Pennsylvania.

This horrible battle took place between July 1 and 3, 1863. When the Confederates arrived on the 1st, in Gettysburg, Confederate leader Richard S. "Old Baldy" Ewell had failed to take Cemetery Hill, because he did not want to take control of the hill unless it was unoccupied. The Confederate army would pay dearly for the unfortunate missed opportunity (pg 106-7, #13). Ewell discovered Cemetery Hill on July 1st, 1863, located at Evergreen Cemetery. Ewell had the opportunity to seize two hills that some said he should have claimed during the course of the battle (pg 174, #13). He was ordered to support General James Longstreet's attack on Cemetery Hill, on which the Union's 50 guns simultaneously attacked Confederate forces; 15,000 men fell in three hours between 4 and 7 p.m. on July 2nd. An additional 1,500 men would fall in evening battles at Cemetery Hill and Culp's Hill.

Meanwhile, Union Commander Andrew Humphreys had lost 1,500 of his 4,000 troops during a Confederate attack at Peach Orchard (pg 167, #13). Colonel William Colvill led his 1st Minnesota Regiment in a charge against a Confederate brigade at Cemetery Ridge. Of the 262 Union soldiers who made the charge, 215 died (pg 169, #13). The town of Carlisle was occupied by Union forces when Ewell and "Jeb" Stuart reached the city with Confederate forces (pg 199, #13). Stuart found Lee on Seminary Ridge; he was not happy with the situation. Lee then ordered attacks by Ewell and Longstreet at first daylight on July 3rd. Commander Alpheus Williams, a 52-year-old Union Division Commander, had no West Point education but was nevertheless considered a good field commander. He had just as many troops as the Confederate leader "Allegheny" Johnson had (pg 202, #13). Yet more importantly, Williams had much more ammunition than the Confederate army had (pg 202, #13). Lee wanted to assault the Union forces on both Union flanks and wanted to use 150,000 men, though he had far less (pg 203, #13).

Longstreet, however, saw disaster in this plan, and he informed Lee of it. At 4:30 a.m., Union guns from Maryland pounded Confederate Maryland forces at Culp's Hill (pg 205, #13). In response to this, the Confederates made two attacks, both of which were repulsed by Union forces, and 78 Confederate Maryland soldiers were taken prisoner. Longstreet's morning attack on the third of July had been delayed, which hurt the Confederate forces (pg 208, #13). The grand attack was a Confederate assault that took place from 1 to 3 p.m. on July 3rd, 1863, which would come to be known as "Pickett's Charge," named after Confederate Major General George Pickett. It is reported and believed that Pickett's Charge was when the Confederacy became mortally wounded.

Yet the last two years of the war saw more fatalities than the first two (pg 213, #13). The purpose of Pickett's Charge was to use the Confederacy's 160-gun bombardment to advance between Union armies without having to endure big federal guns in the process. Lee then wanted to attack the single most vulnerable point of the line, in theory. Unfortunately for General Lee, he had already used a lot of the ammunition that could have been used in another manner.

Longstreet saw that the Confederate forces were almost out of ammunition and thought it unwise to attack the Union army (pg 220, #13). Consequently, the Confederate forces failed to neutralize the federal gunners, and the Union artillery remained in their calculated positions.

They were stationed in order to take out the Confederate assault (pg 226, #13). At 1,220 yards from the enemy, the Union guns opened fire, reducing the number of Confederate soldiers exercising the grand assault by a quarter. In spite of these numbers, the Confederates marched towards the federal line with hundreds of their fallen soldiers behind them.

Of the 12,500 Confederate soldiers who had originally made the assault, only half made it across Emmitsburg Road to ascend Cemetery Ridge (pg 234, #13). It was there, along Emmitsburg Road, that Pickett's forces were overwhelmed (pg 235, #13).
They were assaulted to the right by the Vermont Brigade, led by Brigadier General George Stannard. The Confederate forces entered a box, trapped on all sides (pg 239, #13). Consequently, the Confederate soldiers surrendered and the Confederate flag was captured (pg 241, #13).

The situation was so bad that Pickett reported to General Lee that he had no division left to counter attack. Also on that day, General George Custer led the First Michigan Battalion on a charge at Cress Ridge, which resulted in 254 Union soldier losses, compared to 181 Confederate fatalities (pg 249, #13). Judson Kilpatrick ordered a foolish and unnecessary charge against the Confederate right wing, resulting in many deaths. This would also accomplish nothing militarily (pg 255, #13).

The final day of battle, July 3rd, 1863, was a disaster for the Confederate army (pg 259, #13): 6,800 Confederate soldiers were left behind. Accepting defeat, General Lee abandoned the town of Gettysburg on July 4th, 1863. Lee accepted his role in the defeat (pg 276, #13). As a result of this loss, Lee headed with his forces to Virginia. General George Meade of the Union chased Lee into the south but failed to capture him and his army, due to Meade being reluctant to lose any more men. Thus was forfeited another golden opportunity to capture General Lee and terminate the war, once and for all.

President Lincoln honored the memories of the fallen on the battlefield four months later by declaring that every man who died there, on both sides, gave their lives for freedom (pg 292, #13). Lincoln did this on November 19th, 1863. However, the failure of Meade to pursue Lee into Virginia drew the wrath of Lincoln, who stated, "Meade and his army had expanded their skill and toil and blood up to the ripe harvest, then allowed it to go to waste!" (pg 450, #4). The Union victories at Gettysburg and Vicksburg were practically simultaneous in July 1863.

Grant was rewarded for his victories by President Lincoln with a promotion to major general (pg 256, #11). In fact, Lincoln rated Grant as his best general. Lincoln also had the wisdom to note that a major factor in the Union victories at Vicksburg was the employment of Negro troops. Comparatively, the Confederate army did not use Negroes as troops, and their power was finally beginning to disintegrate. Perhaps one of the biggest contributors to the end of the Confederate cause was the death of

"Stonewall" Jackson shortly after the Battle of Chancellorsville, on May 10th, 1863. Had Jackson lived, perhaps he would not have made the same mistakes as Ewell did for Lee at Gettysburg. This scenario could have possibly changed the outcome of the Gettysburg battle and that of the war.

After Gettysburg, Lincoln appointed military governors for the territory recaptured by the Union army, including Tennessee, Louisiana, Arkansas, and North Carolina, while the war was in progress. This tactic was implemented with the intention of quickly restoring all southern states once the Union victory was guaranteed (pg 469, #4). Lincoln, as commander-in-chief, made the point of blaming the cause of the conflict on individual southerners and not on the southern states as a whole. These individuals were fighting against a united, national, federal government. This political tactic gained Lincoln support from both the radical Republicans and the war Democrats (pg 476, #4).

Yet in New York City, civil problems were starting to explode. There were draft riots over the war. The Irish were afraid that their jobs would go to newly freed blacks, so a race war erupted in the streets of New York City (#9). Contrary to the white bigotry, however, African-American soldiers joined both the Union Army and Navy and served in a total of 166 regiments. In total, 178,985 African-American soldiers served in the Civil War, which made up approximately 10% of the Union army (#9).

General Grant pressed through Mississippi and Alabama as Commander of the Armies. General Rosecrans was successful in recapturing Chattanooga, Tennessee. But Rosecrans followed Confederate General Braxton Bragg to Chickamauga Creek, 12 miles south of Chattanooga, and met a reinforced Confederate army backed by General Longstreet. This was on September 19th, 1863. After brutal and intense fighting there, the Union army lost the Battle of Chickamauga Creek, with casualties on both sides estimated at 38,000 men (pg 263, #11). Earlier, the Battle of Shiloh was one of the bloodiest of the entire war, with casualties of 13,047 Union soldiers and 10,699 Confederate soldiers. Confederate General Albert Sydney Johnston also perished in this battle. Nevertheless, it was considered a Union victory.

The Battle of Mechanicsville was also a Union victory and considered a big win. Because of these victories, Hooker would redeem himself on November 24th, 1863, when he successfully retook Lookout Mountain by the Tennessee River (#9). The successful Chattanooga campaign ended the Confederate threat west of the Allegheny Mountains for the rest of the war (#9).

To reward his success for the Union cause, President Lincoln promoted Grant to lieutenant general, the first man since George Washington to have that title. The measure was approved by the House and Senate also (pg 286, #11). Grant was thus placed in command of the combined Armies of the United States by President Lincoln on March 9th, 1864 (#9). The

Union plan was that Grant's army would attack Lee's, while General Sherman attacked Joseph Johnston (pg 488, #8). Ironically, if Lincoln had his choice of all generals, he would have appointed Lee as commander of the Army of the Potomac.

The goal of the simultaneous Union attack on Atlanta and Richmond was to capture both southern strongholds (#9). Grant was determined to capture Lee at all costs. This was the opposite of the actions of General McClellan. While Grant had his assignments in the eastern theatre, Grant gave command of the western theatre to General William "Tecumseh" Sherman. Tragedy befell Union forces on April 12th, 1864, at the Ft. Pillow Massacre, led by Confederate General Nathan Forrest. After 231 black and white Union soldiers surrendered to Forrest, all 231 were murdered. When the war ended, Forrest would go on to become the first Grand Wizard of the Ku Klux Klan in 1866 (#9).

The ferocious battles would continue in 1864, with General Grant pursuing Lee. In the Battle of the Wilderness, which took place on May 5th and 6th,1864 , both Generals Grant and Meade sustained heavy losses due to the guerilla warfare tactics used by the Confederate troops (#9). When the battle was over, both sides had heavy losses and many wounded. For the Union, total casualties were 17,666 men: 2,246 killed, 12,037 wounded, and 3,383 captured or missing. Lee's forces suffered 11,000 casualties.

Grant was relentless in his pursuit of Lee, but he sustained heavy casualties in the process. Grant was eager to end the war, once and for all. Grant marched his troops to Chancellorsville, Virginia, where a monstrous battle took place. Union forces were intercepted there while crossing the Rapidan River, and 17,000 men were lost over two days of fighting in May 1864. This tragic loss of life for the Union caused Grant to weep in his tent (#3). The fighting continued in Spotsylvania County, Virginia, another ferocious battle. As terrible as this battle was for both sides, it ended in an essential draw. Though Lee and his army fought hard against the Union, they were outnumbered and were wearing down. Lee had not experienced a Union general like Grant. Grant was intent on victory, no matter what. Yet that passion for Union victory came at an extremely heavy price.

After two and a half weeks of battle in May 1864, Grant's army sustained 33,000 deaths, averaging about 2,000 deaths a day. These fatalities included the Battle of Spotsylvania, where Lee and Grant met against each other for the first time in May 1864. This terrible battle resulted in 17,000 casualties for the Union, and between 9,000 and 10,000 for the Confederacy (pg 495, #8). However, a catastrophic loss of command for the Confederates was the death of General "Jeb" Stuart at the Battle of Yellow Tavern, on May 11th, 1864. Stuart was second in importance to the Confederate military, behind only Lee himself (#9). Grant was intent on pursuing Lee on his way to Richmond, and on ending the war. In fact, Grant felt pressure from the Union to end the war before the presidential election of 1864, which would have been a great benefit to Lincoln.

It was this desire of winning the war that led Grant into one of his most costly defeats in early June 1864, where the two armies met in a brutal battle called Cold Harbor. This mistake by General Grant was partially due to his misjudging the strength of his opponents, who had set a trap in order to split Grant's army in two. Grant's military objective was to force Lee's army to fight out in the open, and not in the woods, where Grant believed that he held the numerical advantage. However, unknown to Grant, Lee had already entrenched the area of Cold Harbor so that the Confederates were successfully disguised by logs to hide their superior position. For the Connecticut Legion, this was their first experience in serious combat. Most of the Union army at Cold Harbor had no experience with entrenched warfare.

Grant had to postpone the Union attack on Cold Harbor until June 2nd because the Union army had gotten lost in the woods. Then, the Union attack was delayed from 5 p.m. on June 2nd until 4:30 a.m. on June 3rd, which gave the Confederates time to dig fortified trenches. Lee also knew the terrain of the land better than Grant (pg 177, #7). This delay in attacking only helped the Confederates. When Union troops went into battle at Cold Harbor, they were struck by catastrophic defeat: 7,000 Union troops were killed in the first hour (#9). Maryland and Florida military units for the Confederates trapped the routed Union soldiers as they fled; Union forces suffered 12,700 casualties during the tragedy of Cold Harbor. Lee's army had achieved a total victory, but he admitted to being lucky by only losing 1,500 men.

The terrible Cold Harbor battle was characterized by those who were there as "the valley of the shadow of death." In response to the Union defeat at Cold Harbor, the Union troops dug trenches for a possible Confederate advance, which never came. Around June 6th, 1864, a truce was called for by both sides, so that the dead and wounded could be cleared. Grant, emotionally devastated but not deterred by the loss at Cold Harbor, continued his pursuit of Lee to Richmond in an effort to win the war.

The Union armies learned that trench warfare was much more effective than standing up and shooting at one another. Grant had learned a lot from his loss at Cold Harbor. This would serve Grant well when he achieved victory against the Confederacy 10 months later. Although Lincoln was depressed from the human losses, he believed wholeheartedly in General Grant, because he would not retreat. After the Union's disastrous defeat at Cold Harbor, Grant's army continued to cross the James River, where he could cut Lee's supply line to Richmond. Lee retreated to Richmond because he deemed it to be the biggest vulnerability to the Confederacy (pg 370, #11).

Grant crossed the James River with complete success (pg 373, #11). Grant's army continued to attack the Chickahominy River towards the city of Petersburg. Petersburg was a critical rail junction that supplied Richmond and lay 22 miles south. In June 1864, the Siege of Petersburg began, with 16,000 Union troops attacking 3,000 Confederate troops, yet the Union was repulsed twice after a fierce siege that proved prolonged and costly (#9). On the third invasion, and after heavy bombardments on the city, Petersburg did fall to Grant. There would be heavy losses in casualties for both sides at Petersburg.

On July 17th, 1864, Jefferson Davis replaced Joseph Johnston with John Hood in order to keep General William Tecumseh Sherman from Atlanta (#9). Despite this, Atlanta ended up in flames, and General Sherman was successful in recapturing the city. Sherman ordered the complete destruction of the city (#9). Ironically, Lincoln's former commander, General George McClellan, became his Democratic presidential opponent in the 1864 presidential election. During this time, General Philip Sheridan successfully repulsed Confederate forces attempting to invade Washington, D.C., with heavy losses sustained by the Confederates.

Lincoln himself was shot at by Confederate forces in the Battle of Ft. Stevens, one of the defensive forts outside D.C. Lincoln was reviewing the defenses on July 11, 1864 when the Confederates attacked, with him becoming the only U.S. President to face enemy fire while in office. Future Supreme Court Justice Oliver Wendell Holmes was stationed at the fort and reportedly shouted at Lincoln to "get down, you fool!" during the incident.

The death of Chief Justice Roger Taney on the Supreme Court led Lincoln to appoint Salmon Chase to that position, because he and Lincoln were at odds since Chase had resigned his Cabinet position. Democrat Andrew Johnson replaced Hannibal Hamlin on the 1864 Lincoln presidential ticket as the vice presidential candidate, because of his strength in southern and western states. This was true also because of Hamlin's lack of popularity in the New England states (pg 505-6, #4). Johnson also stuck with Lincoln when his state seceded from the Union.

At the end of the campaign in 1864, Lincoln overwhelmingly won the presidency a second time. African-Americans and Protestant religious groups helped Lincoln in the voting booth. Meanwhile, Confederate General John Hood tried to divert Sherman from his massive raids on Georgia and South Carolina by attacking Union forces in Tennessee. Yet, Major General George Thomas effectively neutralized Hood at the Battle of Nashville (#9). At Cedar Creek, Union General Philip Sheridan fought Confederate General Jubal Early, with 5,665 killed, wounded, or missing for the Union out of 30,829 engaged and 2,910 killed, wounded, or missing out of the 18,410 engaged, respectively. As a result, Sheridan burned all of the crops in the Shenandoah to hurt the Confederacy (#9). Atlanta surrendered to Sherman by September 1864.

Lincoln, after his second successful presidential election, immediately tried to put aside political differences with the Democrats so that they could both unite in order to save the Union. The death threats continued to escalate against Lincoln, but he paid them less than their deserved attention (pg 548, #4). After Sherman had successfully captured Atlanta, Lincoln happily smelled victory in the air. In a blaze of political glory, on January 31st, 1865, Lincoln officially signed the 13th Amendment to the U.S. Constitution, which banned slavery in America.

Meanwhile, in order to win control of Petersburg, Grant and Sheridan wanted control of Five Forks, which was a junction vital to the Confederate army line of supply. Lee knew that if Five Forks was taken by the Union, then he would not only be cut off from Johnston but also would be forced to evacuate Petersburg, as well as Richmond (#9). The Confederacy was beaten by Sheridan, and Pickett lost 5,000 men, who were captured by the Union. Five Forks is described as "the Waterloo of the Confederacy" (#9). President Lincoln, who was already in reconstruction mode before the official end of the war, offered federal money to compensate slave owners for their losses, estimated at $400 million (pg 560, #4).

Lincoln's main objective was "to secure not merely peace, but reconciliation" (pg 574, #4). Back on the battlefield, Grant finally broke through Lee's lines after nine months of battle at Petersburg on April 2nd, 1865. This battle resulted in the death of Confederate General Ambrose Powell Hill. As a consequence, the Confederate capital of Richmond was evacuated on April 2nd, 1865, and relocated to Danville, Virginia (#9). As a result, Union troops quickly occupied Richmond. Lincoln met Grant in Petersburg in order to congratulate him (#9).

Confederate General Jubal Early's initial assault on Washington was repulsed by Philip Sheridan, who laid waste to the Shenandoah Valley. President Lincoln, the first president to be elected to two terms since Andrew Jackson, went to Richmond to go to Jefferson Davis's office and sit in his presidential chair (#9). Lee still believed that he had a chance of a negotiated settlement for the Confederacy, but he knew he was beaten (#9). Lee went to Appomattox Court House after fleeing Petersburg (#9). Lee saw that he was surrounded and beaten. Instead of continuing the bloodshed with guerilla warfare from the hills for years and years, Lee chose to completely surrender in order to heal the country. Lee did so with his 28,356 soldiers to Grant at the Appomattox Court House on April 9th, 1865. By April 10th, the war was over. Despite this, technically the last fighting unit for the Confederacy was led by Edmund Kirby Smith in Texas, who surrendered on May 26th, 1865. The last general to surrender was Cherokee leader and Confederate Brigadier General Stand Watie on June 23rd, 1865 (#9).

According to the rules of surrender, the Confederates could just put down their arms and go home, obeying the laws of the United States. Grant was generous enough to Lee to let the Confederate men keep their horses. Robert Lee would go on to become president of Washington and Lee College. He died in 1870. After Lincoln saw that African-Americans played a huge role in the Union victory, he said that they belong in our society as a "permanent part of the social American fabric" (pg 583, #4).

John Wilkes Booth was in attendance at Lincoln's second Inaugural Address and was in striking distance of killing the president. Booth loathed the president, was a southern sympathizer, and was a well-known actor in the theatre circle. The Confederate surrender on April 9th, 1865 triggered Booth to act of his own accord as a Confederate agent. Lincoln, in the meantime, wanted to bring back the Confederate states on the most generous of terms in order to achieve a speedy restoration (pg 590, #4). Lincoln was truly a different man after the Union war victory. Senator James Harlan said of Lincoln, "His whole appearance, poise, and bearing had marvelously changed. He seemed the very personification of supreme satisfaction" (pg 593, #4).

Lincoln decided to go to a performance at Ford's Theatre to see "Our American Cousin," against advice from his Cabinet not to attend. This was due to possible threats on Lincoln's life. However, Lincoln chose to attend the play to avoid talking to endless visitors at the White House. Booth employed two other assassins, George Atzerodt and Lewis Paine, to kill Vice President Johnson, who was staying at Kirkland House, and Paine was also to kill Seward, who was staying at home. Booth had no trouble getting into Ford's Theatre, because he was a well-known actor there, so it would have been common to see him at a performance (pg 597, #4).

Booth was granted access to Lincoln's presidential balcony that overlooked the stage. Booth shot the President at approximately 10:13 p.m. (pg 597, #4). Atzerodt did not try to kill Vice President Johnson, but Paine did try to kill Seward by breaking into his home and stabbing him. President Lincoln was dead by 7:22 a .m. on April 15th, 1865, as the final symbolic casualty of the most tumultuous war in American history. Lincoln gave his life to protect the freedom and liberty that all of us enjoy and take for granted today. For his great sacrifice for democracy and freedom itself, we can always be grateful to Abraham Lincoln. Ulysses S. Grant called Lincoln "the greatest man I ever knew."

In the Civil War, from 1861 to 1865, there were 10,600 engagements. In the Union army, of the 2.1 million men who served, 360,000 died: 110,000 men died in battle, while the rest either died of disease or accidents. In addition, 275,000 men were wounded in battle for the Union. The death rate for Union soldiers was one in six, or 17%. Of the approximately 750,00 to 850,000 men who served for the Confederacy, 260,000 died: 100,000 in battle and 160,000 from disease. In addition,

200,000 Confederate soldiers were wounded. The death rate for Confederate soldiers was one in three.

The final fatality count was 620,000 for both sides of this awful war, and 1.1 million on both sides were wounded. All the wars in this nation's history combined do not add up to the casualty list of the Civil War. Thirty-four percent of the men from the Confederacy served, while 17% of the men in the Union served; 65 Union generals died, as opposed to 92 Confederate generals.

On the economic side, the federal budget for 1860 was about $63 million. In 1865, at the war's end, that increased to $1.3 billion, an increase of 200-fold. By 1879, $6 billion were spent in paying the veterans' pensions. Confederate costs of the war were approximately $2 billion. The destruction of the south included two thirds of all wealth being swept away, including slavery. Half of all the southern farming industry was destroyed. In the years 1860-70, northern wealth increased by 50%, while southern wealth decreased by 60%.

Why did the war end the way it did? Why did the Union win? One reason is that the Confederates never became nationalists. They would not relinquish states' rights in favor of the new Confederate union. The Confederates did have a strong union of government, with leaders like Jefferson Davis and Robert Lee, but there were no other opposing political parties. Women, who had no voting rights anyway, gave up on the Confederate cause quicker than did men. The Union ultimately prevailed in the war because of better men and equipment, and the northern people supported the war throughout. The Civil Rights Act of 1866 declared all African-Americans to be citizens of the United States (#9). Despite President Andrew Johnson vetoing the Act, Congress overruled Johnson to make it federal law.

Bibliography
1.*American Heritage: The History of the Battle of Gettysburg*, by Craig L. Symonds (HarperCollins, 2001), disc 1.
2. *Lincoln: A Contemporary Portrait*, by Allan Nevins and Irving Stone (Doubleday, 1962), disc 1.
3. History Channel video: "Civil War Combat: Tragedy at Cold Harbor," disc 6.
4. *Lincoln*, by David Herbert Donald (winner of the Pulitzer Prize) (Simon & Schuster, 1995), disc 2.
5. History Channel video: "Civil War Combat: The Bloody Lane at Antietam," disc 2.
6. Professor Gary Gallagher's lecture on the Civil War, University of Virginia (audiotape at local library).
7. *Grant as Military Commander*, by James Marshall Cornwell (Barnes & Noble Books, 1970, 1995), disc 5.
8. *The Wars of America*, by Robert Leckie.

9. *Encyclopedia of the American Civil War: A Political, Social, and Military History,* by David S. Heidler and Jeane T. Heidler, eds. (ABC-CLIO, 2000).
10. *Antietam: The Photographic Legacy of America's Bloodiest Day,* by William A. Frassanito (Thomas Publications/Gettysburg, 1978), disc 7.
11. *Grant,* by Jean Edward Smith (Simon & Schuster, 2001), disc 4.
12. *Team of Rivals: The Political Genius of Abraham Lincoln,* by Doris Kearns Goodwin (winner of the Pulitzer Prize) (Simon & Schuster, 2006).
13. *Gettysburg,* by Craig Symonds (HarperCollins, 2001).
14. *The Library of Congress Civil War Desk Reference,* by Margaret Wagner, Gary W. Gallagher, and Paul Finkelman (Simon and Schuster, 2002).
15. *The Sable Arm: Black Troops in the Union Army, 1861-1865,* by Dudley Taylor Cornish (University of Kansas Press, 1987).
16. *A Compendium of the War of the Rebellion: From Official Records of the Union and Confederate Armies, Reports of the Adjutant Generals of the Several States, the Army Registers, and Other Reliable Documents and Sources,* by Frederick H. Dyer (T. Yoseloff, 1959).

NOTES ON PROFESSOR GARY GALLAGHER'S LECTURE ON THE CIVIL WAR

There were 10,000 engagements in the war. In the North, of the 2.1 million men who served, 360,000 men died: 110,000 men died in battle, while the rest died of disease or accidents, and 275,000 men were wounded in battle. The death rate for soldiers was 1 in 6, or 17%. For the Confederates, of the approximately 750,000 to 850,000 men who served, 260,000 died: 100,000 men died in battle and 160,000 from disease, and 200,000 additional southern soldiers were wounded. The death rate for the South was 1 in 3.

In total, 620,000 men died on both sides, and 1.1 million men on both sides were wounded. All wars this nation has been in combined do not add up to the casualties of the Civil War. Thirty-four percent of men from the South served in its army, while about 17% of men in the North served. Sixty-five northern generals died, as opposed to 92 southern generals. On the economic side, the federal budget in 1860 was about $63 million. In 1865, at the war's end, the number increased to $1.3 billion, an increase of 200-fold.

By 1879, $6 billion had been spent for paying veterans' pensions. Confederate costs were approximately $2 billion. Two thirds of all wealth was swept away, including profits from slavery, and half of the farming industry was destroyed. In the years 1860-70, northern wealth increased 50%, while southern wealth decreased 60%.

Why did the war end the way it did? Why did the Union win? One reason is that the Confederates never became nationalists. They would not let go of states' rights in favor of their new Union. The Confederates did have strong union of government, with leaders like Jefferson Davis and Robert E. Lee, but no opposing political party system. Women, who had few rights anyway, gave up on the Confederate cause quicker than the men. The North ultimately prevailed with better men and equipment, and the northern people supported the war throughout. President Lincoln explained to the northerners why the war was justified. In the spring of 1863, the Vicksburg campaign brought defeat for the North, but Sherman and Sheridan subsequently brought victory back to the North.

Chapter 8
The Spanish-American War and President William McKinley's Decisions and Actions During This Conflict

The U.S. military involvement in the Spanish-American War lasted from April 21st to August 12th, 1898. This was a war in which the United States was successful in defeating Spain in the islands of Cuba and the Philippines in less than 4 months. In doing so, the United States freed the Cuban people from 400 years of extremely oppressive Spanish rule (pg 8, #1). Spain once had control of many lands in the New World, yet they had no existing colonial policy, like England once had.

This war, in 1898, fell on the shoulders of then-President of the United States William McKinley, who had served in the Civil War for the Union. McKinley saw firsthand the horrors of war and knew the high price to be paid for freedom. McKinley's predecessor, President Grover Cleveland, wanted a more cautious approach to intervention in the Caribbean than did President McKinley (pg 32, #1). When McKinley was sworn in as president in 1897, he had agreed to give Spain "limited time" to find a solution to the harsh oppression being inflicted on the Cuban people. From the Spanish point of view, it had an interest in keeping Cuba a Spanish colony. However, everything changed when, on February 15th, 1898, two suspected mines blew up the *U.S.S. Maine* in Havana Harbor, Cuba, while it was docked. This catastrophe killed 266 of the 350 crew onboard.

In order to understand the events that led to the destruction of the *U.S.S. Maine* and to the Spanish-American War, in which future President Theodore "Teddy" Roosevelt would serve in battle, I will first examine the life and political philosophy that characterized President William McKinley. I will describe how McKinley handled this particular crisis and how he and his political advisors achieved military victory, and territorial gain in a relatively quick campaign.

Another very important military and political figure emerged as a hero for his actions in this war, and that was Teddy Roosevelt. Roosevelt, who had a Cabinet position at the time of the war as Assistant Secretary of the Navy, actually left his position in government to enlist in the army as a lieutenant colonel, and produced a famous victory with his gang of troops called the Rough Riders. The Rough Riders achieved a very important victory at the Battle of San Juan Hill. Because of Roosevelt's personal accomplishments in this war, he was subsequently elected governor of New York in November 1898, and later vice president on the McKinley Republican ticket in 1900. In order to understand the successful liberation of Cuba from Spain, it is important to start at the beginning of this conflict, based on political and historical events at the very end of the 19th century.

The war with Spain was the first American war with a European power since the War of 1812 (pg 91, #2). As this war was President McKinley's decision, I will analyze his life history before and while he was president. William McKinley was born on January 29th, 1843, in Niles, Ohio. He was a Methodist Christian, although he was originally Presbyterian. This flexibility would serve him well later in life by allowing him to be tolerant of other religions.

During the Civil War, McKinley served with the 23rd Ohio Volunteer Infantry from June 1861 to July 1865. During this time, McKinley advanced in rank from private to brevet major. Although he was not wounded in action, McKinley showed courage while under enemy fire in battle. McKinley strongly supported President Lincoln's leadership. While McKinley was in the army for the Union, he served for Colonel Rutherford B. Hayes, who would later become the 19th President of the United States. After the war, McKinley continued to have a great relationship with Hayes.

When McKinley returned home from serving in the war, he took and passed the Ohio bar exam, thereby becoming a lawyer. On January 25th, 1871, McKinley married Ida Sayton from Canton. They would have 2 daughters, Katie and Ida McKinley, who would both tragically die during their infancies. McKinley's first legal job was as a Stark County, Ohio prosecutor. This job would elevate him to his eventual election to the U.S. Congress in 1876. McKinley was then appointed Chairman of the Ways and Means Committee, and had very strong labor support while he was in Congress (pg 32, #2).

McKinley was also known as a friend of enterprise, industry, and tariff protection. For example, McKinley supported the Sherman Antitrust Act of 1890, which protected consumers from unfair price gouging that enriched companies at the expense of the consumer. Ironically, McKinley was against the Republican party on the issues of unions, which McKinley favored and the party did not. While McKinley was friends with Hayes, the two men disagreed on issues such as Hayes's policy on "open reformism." Hayes was against money's big influence on the government (pg 41, #2). Despite these disagreements between Hayes and McKinley, the two men agreed on not sending federal soldiers to the Ohio railroad strike in order to quell the protesters.

114

McKinley also firmly believed in women's rights, and he wanted to improve them however he could (pg 46, #2). As a result of his pro–women's rights platform, McKinley received honorary degrees from two women's colleges: Smith College and Mt. Holyoke in Massachusetts.

McKinley served in his House seat from 1877 to 1883, and then lost his seat for one term. McKinley regained this seat in 1884 and would keep it until 1891. In the years 1889 and 1890, the United States saw the addition of 6 states to the Union: (1) North Dakota, (2) South Dakota, (3) Wyoming, (4) Montana, (5) Idaho, and (6) Utah (pg 63, #2). These states would later serve as Republican strongholds. In the election of 1892, McKinley gave up his seat to run for the governorship of Ohio, which he won. McKinley's leadership and prominence got him noticed in the Republican party.

During President Cleveland's second term in office, there was a big economic depression in the United Sates that lasted from 1893 to 1897. These financial dilemmas hurt the Democrats (pg 67, #2). McKinley started to consider becoming a candidate in the 1896 presidential election. His Republican peers approved. At the Republican convention held in St. Louis in June 1896, McKinley won the Republican nomination on the first ballot.

He campaigned as a candidate in favor of labor, tariffs, the acquisition of Hawaii and part of the Danish West Indies, and women's rights: "equal pay for equal work." His Democratic opponent was William Jennings Bryan, who was a lawyer from Illinois who later became a Congressman from Nebraska. Bryan was a champion of the "free silver" movement and federal income tax (pg 360, #7). He scorned the gold standard of the American economy.

Some Democrats split with Bryan and formed the Eastern Democrats, who nominated John Palmer of Illinois as their candidate. They called themselves the Gold or National Democrats, while Silver Republicans supported Bryan's nomination (pg 361, #7). To McKinley's advantage, he was praised for being the most pro-labor Republican since Abe Lincoln (pg 77, #7). McKinley's campaign manager, Mark Hanna, campaigned vigorously for him and raised $13 million from other industrialists, who were wary of Bryan's populism. McKinley also received valuable support at the polls from Civil War veterans. In fact, McKinley was the last president who served in that war.

On election night, in November 1896, McKinley defeated Bryan by a tally of 7,035,638 votes (51%) to Bryan's 6,467,946 (47%). After his defeat to McKinley, Bryan retained control of the Democratic party and won the presidential nomination again in 1900 (pg 360, #7). When William McKinley was inaugurated on March 4th, 1897, there was a bit of ironic foreshadowing in McKinley's speech when he said, "We want no wars of conquest, we must avoid the temptation of territorial aggression. War should never be entered upon until every agency of peace has failed. Peace is preferable to war in almost every contingency. Arbitration is the true method of settlement of international, as well as local, or individual differences" (pg 363, #7).

President McKinley knew a great deal about global relations from his days in Congress. Garret Hobart of New Jersey was his vice president for his first term. McKinley's presidency was the first to have use of both telephone and telegraph communications. Meanwhile, Cuba wanted to gain independence from Spain, which was brutally oppressive to the Cuban people. By mid-1897, at least a third of the Cuban population had died as the result of Spanish rule (pg 20, #1). General Valeriano Weyler of Spain was responsible for horrible cruelties to the Cubans.

The United States became involved in this conflict when, in October 1873, 53 U.S. sailors onboard the ship *Virginius* were taken prisoner and massacred by the Spanish. Yet Spain paid money to the United States for this crime, hoping to avoid war between the two countries (pg 79, #4). This act delayed the inevitable war between the two countries, though tensions were mounting.

Spain and Cuba had previously fought a war with each other, called the Ten Years' War. In that conflict, Spain defeated Cuba. The treaty that ended the war, called the Pact of Zanjón, was a shameful document that heavily favored Spain. Thereafter, Spanish rule of Cuba was described as cruel and horrible. By the time of the conflict in 1898, the Spanish had broken every rule of the 1878 Pact of Zanjón.

The Cuban people had no rights of free speech, press, or religion. There was also no security of persons or property, as was discovered by a young Winston Churchill (pg 53, #1). During this time, Valeriano Weyler y Nicolau became the new Spanish dictator of Cuba (pg 54, #1). As a precaution of this event, Teddy Roosevelt, who was then Assistant Secretary of the Navy, urged President McKinley to build up the U.S. Navy for insurance reasons in case it was needed for war with Spain. Weyler, once in power in Cuba, issued harsh *bandos*, or decrees, that forced Cuban peasants to move to the cities and give up their farmland. There, in the cities, the peasants would often starve or catch disease.

The reason Weyler gave for this cruel treatment of the Cuban peasants was to rob the rebels of their support base, guerrilla recruits, food, and military intelligence. General Weyler was a cruel man. The situation was described as butchery of the Cuban people by the Spanish. Then Assistant Secretary of the Navy Roosevelt recommended to President McKinley that he build up the U.S. Navy because the navy was not that strong in early 1898.Teddy Roosevelt then said, "To be prepared for war is the most effective means to promote the peace" (pg 593, #5). However, the U.S. Navy did have one tremendous weapon in its arsenal: the *U.S.S. Maine*.

The *Maine* was the largest ship ever built for its time. Its plans for construction were approved in November 1887. By November 18th, 1890, the ship had been completed (pg 74, #1). President McKinley and the

Spanish Ambassador, Enrique Dupuy de Lome, had met in Washington to attempt to avoid war between the two countries, but the talks were described as tense. As a result, McKinley sent the mighty *U.S.S. Maine* to Cuba to protect U.S. citizens who were already stationed there (pg 97, #4).

The situation escalated when a letter was discovered by William Randolph Hearst that exposed the Spanish prime minister's comments about President McKinley: "weak and a bidder for the admiration of the crowd, besides being a would-be politician, who tries to leave a door open behind himself while keeping on good terms with the jingoes of his party." This statement triggered a fresh outpouring of anti-Spanish sentiment. The reasons for the United States to go to war with Spain started to increase in number:

1.   Reports estimating that as many as 200,000 men, women, and children were exterminated by Spain. This was called a "colossal crime" against the Cuban people

2.   Eagerness for conflict on the part of many Americans

3.   Sympathy for the Cubans, who struggled for independence from Spain

4.   Anger at a European power that tried desperately to maintain its colonies in the Western hemisphere

McKinley still tried to avoid war with Spain by diplomatic methods, and by urging Spain to treat the Cubans more humanely. But these efforts were to no avail. Consequently, everything changed between Spain and America on February 15th, 1898. On that day, two explosions sank the *U.S.S. Maine*, which was stationed at Havana Harbor. This incident happened at approximately 9:40 p.m.; 266 of the crew of 350 sailors onboard the *Maine* were killed. The attack caught the American government and President McKinley completely by surprise. In fact, McKinley was deeply upset by the news of the destruction of the *Maine*, which he had learned from Secretary of the Navy John Long (pg 108, #4).

Because there was confusion as to the cause of this explosion, McKinley did not immediately declare war on Spain. McKinley wanted a full investigation as to the exact cause of the explosion before he would decide whether to declare war on Spain, despite urging from Assistant Secretary of the Navy Roosevelt to do so; Roosevelt was convinced that Spain was guilty of the *Maine*'s destruction. In spite of President McKinley's desire to avoid war with Spain, he nevertheless asked Congress for $50 million for military build-up for war with Spain should that happen, and Congress approved (pg 632, #5).

McKinley decided that the United States would conduct a separate investigation of the *Maine*'s destruction, apart from the investigation that Spain would conduct. This decision by McKinley was influenced by Roosevelt, although it was criticized by Secretary Long. There was speculation that the explosion of the *Maine* could have been caused by internal combustion and not by a mine or a torpedo (pg 123, #1).

However, after a Congressional investigation, it was concluded that a Spanish torpedo or mine must have been the cause (pg 119, #4). Yet the cause of the explosion of the *Maine*, to this day, has never been officially determined.

At the time, however, many American citizens were convinced that Spanish mines had destroyed the *Maine*, and they composed the war cry, "Remember the *Maine*, to hell with Spain!" (pg 365, #7). After McKinley saw for himself that the evidence suggested that a Spanish mine had most likely been responsible for the destruction of the *Maine*, and not steam-room overheat, as some suggested, he then began to lean more towards declaring war on Spain. Adding to this argument were McKinley's failed attempts to purchase Cuba from Spain and Spain's resistance to agree to a financial package for the United States to compensate for the destruction of the *Maine*. McKinley then decided to officially begin war preparations. McKinley sent Senator Redfield Proctor of Vermont to Cuba on a fact-finding mission to ascertain the actual conditions there, and Senator Proctor confirmed Spain's horrible atrocities committed against the Cuban people. This was the deciding factor to go to war with Spain from McKinley's point of view. Though McKinley was warned that the economic hit of 1893 would make war too costly, and religious leaders advised him against declaring war, he had the astute ability to read the feelings of the public well (pg 115, #4).

The U.S. Naval Court of Inquiry found Spain guilty of the destruction of the *Maine* in March 1898 (pg 166, #1). Yet the Spanish report, as conducted by Spanish Captain Pedro del Peral, claimed that the explosion of the Maine was caused internally, not externally (pg 174, #1). Captain Charles Dwight Sigsbee of the *Maine*, who survived the explosion, was certainly convinced otherwise.

As a result, on Monday, April 11th, 1898, President McKinley dismissed Spain's offer to suspend hostilities in Cuba. On April 20th, McKinley signed the Cuban resolution, saying that Spain had to leave Cuba or there would be war (pg 202, #1). On April 21st, 1898, McKinley officially asked Congress for a declaration of war against Spain, in order to eliminate Spanish persecution of the Cuban people by means of starvation, murder, and misery (pg 121, #4). In his words, McKinley asked Congress for the authorization "to take measures to secure a full and final termination of hostilities between the government of Spain and the people of Cuba, and to secure in the island the establishment of a stable government, and use the naval forces of the United States as may be necessary for these purposes" (pg 365, #3).

Even though McKinley had upset some war hawks by delaying America's entry into the war after the sinking of the *Maine*, he did so to evacuate U.S. citizens from Cuba (pg 95, #2). McKinley's war declaration passed in both the House and the Senate by a wide margin (pg 366, #3). McKinley asked both Congress and the general public for 125,000 volunteer soldiers for this campaign against Spain (pg 206, #1). He felt the urgent need for soldiers because there were only 25,000 men in the U.S. Armed Forces at the time. Also, McKinley believed that this conflict might be useful in healing the wounds of the Civil War, with northern and southern soldiers fighting side by side against a common enemy (pg 144, #4). Yet the main reason for the war in Cuba was to free the Cuban people from Spanish aggression and oppression.

During this time, Admiral George Dewey took his naval squad to Hong Kong in preparation for battle against Spain in the Philippines. In fact, the conditions in the Philippines and Cuba were very similar (pg 216, #1). Admiral Dewey, who was sixty years old at the time of the Spanish-American War, had previously served during the Civil War for the Union under David Farragut (pg 217, #1). Dewey and Teddy Roosevelt were friends (pg 218, #1). Roosevelt, quite shockingly, quit his job as Assistant Secretary of the Navy to enlist in the army, where he became a lieutenant colonel in charge of the "Rough Riders." The Rough Riders were an assembly of professional soldiers who had participated in campaigns against Native Americans. President McKinley 's plan was to overwhelm the Spanish Army into giving up quickly, and return rule to the liberated Cuban people once the military objective had been met (pg 165, #4).

Before the invasion of Manila, the Spanish military tortured and killed American sympathizers located there (pg 218, #1). Aware of this fact, Admiral Dewey got his men enraged by reading them statements made by the Spanish military leaders, calling U.S. soldiers scoundrels, thieves, murderers, violators of women, and destroyers of religion. This succeeded in firing up his men for the battle that would become the Battle of Manila Bay. This battle, led by Admiral George Dewey, involved six U.S. ships: *Boston, Concord, Petrel, Raleigh, Baltimore,* and *Olympia*, which attacked Spanish strongholds in Manila while defending against 10 Spanish warships. These Spanish ships were led by Admiral Patricio Montojo (pg 366, #3). Yet these six U.S. warships performed exceptionally well in battle (pg 231, #1). The Battle of Manila Bay was fought on May 1st, 1898, beginning approximately at dawn and ending around noon. Eight of the 10 Spanish warships were sunk, and 381 Spanish soldiers were killed or wounded. There were no American fatalities, and just six wounded.

Manila surrendered to U.S. forces that day. The Spanish fleet that Dewey faced was described as extraordinarily inept (pg 349, #8). The United States would eventually control all of the Philippine Islands. Meanwhile, Admiral William T. Sampson followed the Spanish Navy to Havana for impending battle. His counterpart, General Pascual Cervera, wanted to strike Key West (pg 247, #1), and the United States wanted Havana Harbor blockaded (pg 254, #1). Admiral Sampson and the ship *New York* met with the ship *Oregon* in Havana Harbor (pg 270, #1).

Because the *Oregon* had taken so long to get to Cuba from the Pacific,

Teddy Roosevelt saw the need for what would eventually become the Panama Canal, during his Administration in 1904.Though the Spanish Navy went on the offensive against the United States on May 28th, the U.S. military had their ships successfully closed off in Santiago during June. The U.S. Army could now begin its offensive. The Rough Riders that Roosevelt commanded consisted of woodsmen, farmers, cowboys, blacksmiths, and mountain men. They would include Dudley Dean, who was Harvard's greatest quarterback, and Joseph Sampson Stevens, who was the world's greatest polo player (pg 274, #1). Roosevelt even got into some trouble with his superiors by buying his troops beer before taking them into warfare, in an effort to bond with them and show that he was one of them. Roosevelt believed in racial equality among his soldiers, and he treated them equally (pg 209, #4).

When Roosevelt was made a lieutenant colonel, he said, "All men who feel any power or joy in battle know what it is like when the wolf rises in the heart." Roosevelt also stated that, during this time, he was not "in the least sensitive about killing any number of men if there was adequate reason" for doing so. In response to McKinley's plea to the public for volunteer soldiers numbering around 125,000, he got one million volunteers. The opposing Spanish troops were thought to number about 278,000 men (pg 278, #1).

When the U.S. Army landed on Cuban soil, it met very little resistance from the Spanish (pg 177, 1#4). The U.S. bombing of Santiago Harbor started on June 6th, 1898 (pg 287, #1). Captain Sigsbee proposed that the United States should seize Guantanamo Bay, which they did (pg 291, #1). The first land battle occurred at Daiquiri, in which the Americans won in one hour (pg 182, #4). The fighting continued at a place called El Caney, which resulted in intense bloodshed on both sides. Though the U.S. military won this battle, it came at the heavy cost of 205 soldiers killed and 1,080 wounded. The Spanish military lost 215 men and had 376 wounded (pg 198, #4). This amounted to much more than the casualties Lieutenant Colonel Roosevelt suffered in his battles, in which he lost 16 men and had 50 wounded. But the U.S. Army and Navy were clearly winning.

The U.S. Army invaded Las Guasimas, Cuba, next. There was heavy fighting, which cost the lives of six Rough Rider soldiers and wounded an additional 34 (pg 673, #5). Despite this, the United States would still won this battle.

Next came the infamous Battle of San Juan Hill, led by Roosevelt and his Rough Riders. There they would be victorious on July 1st, 1898 (pg 688, #5). The final invasion of Santiago was next. Meanwhile, at sea, Admiral Cervera lost all of his Spanish ships in an ill-advised attack on July 3rd, which killed 323 and wounded 151 out of his 2,200 soldiers. America's losses, in comparison, were one killed and two wounded (pg 359, #8). This was because the U.S. Navy was better equipped and possessed greater technological advancements in its ships compared to the Spanish Navy. Admiral Cervera was captured onboard the destroyed ship *Maria Teresa*, while her captain, Juan B. Lazaga, was killed. The Spanish ship *Furor* was also sunk by the U.S. Navy (pg 350, #1).

McKinley wanted to acquire the Hawaiian islands, because of his own personal belief in "Manifest Destiny" and also because Japan wanted to acquire them. McKinley saw that Hawaii was the halfway point between the United States and the Philippine islands. Therefore, the House of Representatives passed a resolution that provided for the annexation of Hawaii on June 15th, 1898 (pg 294, #1).

The U.S. ships *Oregon, Texas, Iowa,* and *Brooklyn* performed extremely well in defeating the Spanish Navy (pg 351, #1). Ironically, one of the destroyed ships for Spain, the *Vizcaya*, was the ship that supposedly came to visit New York Harbor in peace when the *Maine* was blown up. Even Spain's best ship, the *Cristobal Colon*, ran aground on the beach in defeat. Santiago fell on July 17th, 1898, and Admiral Sampson achieved total victory. The Spanish fleet was utterly destroyed, and Admiral Cervera would later be court-martialed. The American naval force had grown to three times the size of Spain's.

President McKinley instructed Spain that unconditional surrender was the only acceptable solution to end the hostilities between the two nations, which is exactly what he got. When Spain surrendered officially on August 12th, 1898, via the Paris Peace Treaty, the conditions were that Cuba was to be an independent republic, and that the United States would gain control of Guam, Puerto Rico, and (temporarily) the Philippines. The U.S. government would pay the Philippines $20 million for this acquisition (pg 398, #1). However, the fighting there would continue until 1902, and an additional 4,000 American soldiers and 20,000 Filipino soldiers would be killed by July 4th of that year (pg 416, #1).

During the entire Spanish-American War, the U.S. military lost 345 soldiers in combat, and an additional 2,500 perished from disease (pg 98, #2). After beating Spain in the war and obtaining new lands as a result, the United States became a global, imperial power (pg 265, #4). Teddy Roosevelt would return from the war as the most famous man in America (pg 698, #5). He would go on to win the office of governor of New York in November 1898 (pg 714, #5). Yet after this victory, in 1900, when then-Vice President Garret Hobart became ill, it was suggested that Roosevelt become the vice presidential candidate, which he did (pg 741, #5). The ticket would go on to win by beating William Jennings Bryan and Adlai Stevenson, Sr. in 1900.

As for Cuba, it was turned into a democracy after the war, with newly elected Tomás Estrada Palma as its president. But Palma lost control of his government shortly after taking over, due to corruption and scandal, and consequently resigned (pg 419, #1). Roosevelt sent U.S. troops to Cuba to restore order. When the second U.S. occupation of Cuba ended in

January 1909, General José Miguel Gomez became Cuba's second president, but his rule was also full of corruption and violence.

By 1928, General Gerardo Machado y Morales became the full-fledged dictator of Cuba, despite being elected. The Platt Amendment was not renewed in 1934, which would have guaranteed U.S. intervention in preventing insurgent Cuban guerillas from taking power. As a result, Fulgencio Batista y Zaldivar gained control of the Cuban military and had two runs of power until 1959, when, after 2 years of fighting, Fidel Castro Ruz, the son of a Spaniard from Galicia, seized power on January 3rd, 1961. The U.S. then cut diplomatic ties with Cuba.

The Philippines was another U.S. colony that was given self-rule. William Howard Taft would become the first U.S. Governor of the Philippines (pg 421, #1). Manuel Luis Quezon y Molina was elected in 1935 over Emilio Aguinaldo. Japanese troops invaded the Philippines two weeks after the December 7th, 1941 Pearl Harbor attack, and captured the country until the United States liberated it in 1945 (pg 443, #1). On July 4th, 1946, the Republic of the Philippines was declared. Ferdinand Marcos was elected, but he and his regime proved to be very corrupt. Marcos was evicted from power in 1986. In 1988, Corazon Aquino was elected as leader of the Philippines.

After the Spanish-American War, Spain concentrated on domestic affairs instead of on its former colonies. McKinley, at first, favored the idea of occupation of the former Spanish colonies, at least temporarily (pg 265, #4). Yet the situation in America drastically changed on September 6th, 1901, at the Pan-American Exposition in Buffalo, New York. This occurred 100 years and 5 days before the tragedy of September 11th, 2001. A 28-year-old Polish native from Detroit named Leon F. Czolgosz shot McKinley while he pretended to seek a handshake, which mortally wounded the president. McKinley officially died on September 14th, 1901. Teddy Roosevelt then became president. President McKinley's last words were; "It is God's way. His will, not ours, be done" (pg 368, #3).

Czolgosz was tried and found guilty of murdering the president, and he was executed on October 29th, 1901 at Auburn State Prison. President McKinley would come to be remembered as a great diplomat and good economist. McKinley had control of both houses of Congress during his years as president, which helped him implement his policies. McKinley believed in a strong protective tariff system to strengthen America's economy. As a result, McKinley's economic policies would led the United States to a decade of prosperity from 1897 to 1907 (pg 151, #2).

Bibliography
1. *A Ship to Remember: The Maine and the Spanish-American War*, by Michael Blow (William Morrow, 1992), disc 7.
2. *William McKinley*, by Kevin Phillips (Times Books, 2003), disc 6.

3. *The Complete Book of U.S. Presidents*, by William A. Degregario (Barricade Books, 2001).
4. *1898: The Birth of the American Century*, by David Traxel (Vintage Books, 1998), disc 7.
5. *The Rise of Theodore Roosevelt*, by Edmund Morris (winner of the Pulitzer Prize) (Modern Library, 2001), disc 7.
6. *Theodore Rex*, by Edmund Morris (Random House, 2001).
7. *Wars of America*, by Robert Leckie.
8. *U.S. Navy: A 200 Year History*, by Edward Beach (Henry Holt, 1986).

Chapter 9
World War One and President Woodrow Wilson's Response and Actions to a World at War

The global conflict known as World War One took place from the years 1914 to 1918. It was also known as "The Great War." This war was described as so horrible that it became "the war to end all wars." This war would come to involve 36 nations. The nations involved either fought on the Allied side (Britain, France, Russia, and, later, the United States) or on the Axis side (Austria-Hungary, Germany, Bulgaria).

Three new types of warfare became standard in this conflict that brought about mass slaughter of human life on both sides: machine guns, trench warfare, and chemical gases. The enormous prices the troops on both sides paid were based on a strategic game of defense, rather than offense. Trench warfare, for example, would be described as savage and deadly, and included "unspeakable horrors" (#1).

A war that was somewhat of a prelude to the conflict that became World War One was the Austro-Prussian War of 1871, when the newly formed Republic of Germany beat France. In the conflict, France had lost the territory of Alsace-Lorraine to the victorious Germans, and France naturally wanted to re-claim the territory. Also, France had to pay Prussia (Germany's former name) $1 billion dollars for peace (#1).

Back during the Middle Ages, Europe was less a continent of nations, and more of fiefdoms. These fiefdoms were run by feudal rulers. Typically, this meant a monarchical society. The four powers that emerged in Europe during the Renaissance were France, Spain, England, and Italy (#1). The creation of the country Germany from smaller kingdoms was the work of Otto von Bismarck, who was the Prime Minister of Prussia. He was also the First Chancellor of the German Empire (#1).

By 1871, Germany was determined to have more of an influence in Europe. Yet World War One itself began in the unlikely place of Serbia. On June 28th, 1914, a 19-year-old man named Gaurillo Princip assassinated Archduke Ferdinand of Serbia and his wife, the Archduchess Sophie. At the time of the assassination, Serbia hosted a terror organization called Black Hand, which was comprised mostly of youths infiltrating Bosnia to kill (pg 23, #2). As a result of the assassination of the Archduke and his wife, Austria-Hungary declared war on Serbia on July

28th, 1914. This was the beginning of the global catastrophe that would become World War One.

Germany, led by Kaiser Wilhelm the II, backed Austria-Hungary (pg 26, #2). Russia, on the other hand, had mobilized its military in defense of Serbia, because it was bound by a treaty with the country. Consequently, Germany then declared war on Russia, then France declared war on Germany. Germany would declare war on France, also. Germany then declared war on Belgium, and by August 1st, 1914, Germany decided to invade that country. As a result, England declared war on Germany (#1).

Japan joined Britain's cause, and Italy did as well in 1916. At first, this conflict was thought to be a short war (#1). The German war plan, or the Schlieffen Plan, was a formula that was developed well after the end of the Franco-Prussian War of 1870-71. This called for a two-front war: one against Britain and France in the west, and one against Russia in the east.

The French counter-plan was called Plan XVII, which was to stay constantly on the military offensive. Yet what both sides would learn quickly was that by World War One, technology had vastly improved the deadliness of the weapons that were used. At this time, the United States was not involved in the conflict. The German army invaded Belgium on August 1st, 1914. French forces defending Belgium were beaten backwards to within 30 miles of Paris. Yet the German army halted there, out of fear that its supply lines were overextended (#1). Here is where the French army stood tall against its German invaders, by initially halting the German army at the Marne River, and then pushing the Germans back north to the Aisne River. The German halt outside of Paris sacrificed any chance for a quick conclusion to the war (#1). This also established the Western Front of the war, which was principally located in France, Belgium, and part of the Netherlands, while the Eastern Front was in Prussia, Poland, and the borderlands of Russia.

Northern France became the center of intense combat in the war (pg 49, #2). On an average day on the Western Front, 2,533 men would die, another 9,121 would be wounded, and another 1,164 would go missing (#1). These horrible numbers, indicating mass slaughter, were products of the Industrial Revolution, which gave nations identities by mass-producing weapons and munitions, including machine guns, airplanes, and poison gas.

During this time of global turmoil, President Woodrow Wilson was in power and insisted that the United States remain neutral in the dispute. This was done so that the United States could act as chief negotiator between the disputing countries. Wilson believed in four goals to achieve democracy and fairness in the world:

1. an association of nations
2. the guarantee of equal rights for all
3. absolute sanctions against aggressors

4. the manufacture of munitions removed from profit-making corporations and restricted to governments (pg 340, #3)

While Wilson proclaimed the United States' neutrality in Europe, events would force America's entry into this war, on the Allied side. The two main events of provocation against America were the sinking of the ship *Lusitania*, and the Zimmermann note, which will both be described in more detail later in the chapter.

Wilhelm the II, who was Germany's leader, was the last Kaiser (the German word for emperor, based of off Caesar (#1) of that country. His spouse, Augusta Victoria, was the daughter of England's Queen Victoria. Kaiser Wilhelm the II took over power from his father, Wilhelm the I, in 1888, because his father was suffering from cancer. Germany's army was the largest and best equipped of all the European powers of the time (#1). Austria-Hungary's army, however, was poorly led and equipped. The same was true for the Russians, although they had exorbitant numbers of troops. Because there were armies growing by large numbers in Europe, differing nations ended up spoiling for war (#1).

The alliance of England, Russia, and France was officially signed on September 5th, 1914, by all three parties via the Treaty of London. The battles started right away with the First Battle of the Marne, occurring in early September, and it resulted in a French victory. Yet it would come at the steep price of 145,000 lives. This staggering amount of destruction would continue in 1915, with both sides suffering extreme numbers of casualties and deaths.

Yet when the war began in 1914, the Kaiser predicted its early end, due to Germany's previous success against France in the Franco-Prussian War of 1870-71. By World War One, Count Ferdinand von Zeppelin invented the Zeppelin airship for the German military, in order to bomb opposing armies (#1). Howitzers were also introduced by the German military during this war. The French military mistakenly thought that the Germans would invade France through the Ardennes, and not through Belgium. As a result, the German army came within 30 miles of Paris after destructive carnage on both sides (#1).

Next, the French army was driven from the Ardennes to Verdun. British forces were defeated at the Battle of Le Cateau, and sustained 7,800 casualties out of a force of 40,000. As a consequence, British forces had to retreat. To make the situation worse, by the end of August 1914 the French army had lost an astronomical 300,000 men, and the Germans were deep in French territory (#1). To combat this horrible situation, French Commander Joe Joffre decided to defend the Somme River, and wanted Paris to be heavily fortified with French troops.

Once the Western Front of the war was established, it would continue to be a battle line of horrible carnage for four years. On the Eastern Front, in western Russia, the situation was equally bad—so bad that it would eventually cost Czar Nicholas the II his power; the communists took power in 1917, which ended Russia's participation in the war. Yet in Germany, General Helmuth von Moltke was unsuccessful in battle, and he subsequently resigned from command at the end of 1914. He was replaced by Erich von Falkenhayn, German Chief of Staff, in September 1914. At the First Battle of Ypres, concluding at the end of November 1914, the losses on both sides were completely outrageous. For example, the dead, wounded, and missing for English troops totaled 2,368 officers and 35,787 enlisted men. The French army suffered an appalling 50,000 casualties in this battle, while the Germans were hit with an astounding 130,000 casualties. Belgium lost 35% of its entire army at the First Battle of Ypres. The reason for this extensive slaughter was that whenever troops from either side would go on the offensive, they would get slaughtered, because the other side's defensive technology was too strong.

Evidence showing that technology was moving in the direction of massive firepower in combat first surfaced during the Second Boer War in South Africa during 1899-1902, and during the Russo-Japanese War in 1904-05. Despite these lessons from history, World War One military leaders chose to ignore them, which resulted in catastrophic numbers of deaths for the soldiers in combat (#1).

The machine gun can fire up to 450 rounds per minute and can cause extreme carnage to opposing armies. The Industrial Revolution was credited with giving nations the ability to mass-produce not only machine guns, but also airplanes and poison gas. Gas was and is considered a terror weapon (#1). Germany first used chemical weapons to kill thousands of French troops, and to take the rest prisoner (pg 150, #1). To counter this weapon, Britain used its own form of invented gas. Protective masks were developed during this time, yet poison gas would still account for 30,000 combat deaths on both sides.

On December 20th, 1914, the First Battle of Champagne took place, from which the Allies gained 500 yards. However, this gain came at the appalling price of 50,000 Allied troops (#1). Next, the Second Battle of Ypres, in Belgium, took place from April 22nd to May 25th, 1915. This battle became an essential stalemate, with France suffering 100,000 casualties, and Germany suffering 75,000 casualties.

In World War One, Winston Churchill was then the First Lord of the Admiralty, which is equivalent to the Secretary of the Navy in America (#1). Churchill proposed to Britain that the military retake the Dardanelles from Turkey, an Axis power, reopening the Black Sea to Russia (#1). The two Allied goals, proposed by Churchill, were to bomb Constantinople into submission and to retake the Gallipoli Peninsula. However, there were many Turkish soldiers waiting for the invasion, and they were well prepared.

Despite the massive amphibious invasion of Gallipoli by the Allies, they were slaughtered in massive numbers by the Turks. More than 250,000 Allied troops were killed or wounded in this ill-advised attack (#1).

Consequently, the Dardanelle campaign by the Allies was described as a disaster (#1). The Turks did lose 251,000 troops in this battle, but they still controlled the Dardanelles, and Russia was still cut off from any military help from either Britain or France. This fact would eventually cripple the Czar's power in Russia, contributing to his removal from authority by the Bolsheviks.

By the beginning of 1915, the United States was merely an observer to the horror occurring in the European conflict. President Woodrow Wilson, who had no military experience, insisted that his country remain neutral in this dispute and act as chief negotiator between the disputing countries.

Wilson was still in mourning in early 1915 from the death of his wife, First Lady Ellen Wilson, on August 6th. 1914, from Bright's disease (pg 413, #5). But by March 1915, President Wilson had been introduced to his next wife, Edith Bolling Galt. Galt was 43, and a widow herself from the death of her husband, Norman Galt, who was a jeweler from Washington, D.C., who died in 1908 (pg 413, #5). Wilson and Galt were married on December 18, 1915, at Edith's house (pg 357, #3).

During this time, President Wilson's personal feelings about the horrible conflict in Europe were that he was reluctant to go to war, because he felt that domestic policy was too important. Plus, many Americans were from Europe, so Wilson did not want to unnecessarily divide them into factions. This included thousands of Americans who were of German descent (pg 60, #4). However, Britain's and France's economies were very important to the United States, as Wilson acknowledged.

The war drastically changed during 1915, and several acts of terror occurred that sent America on a collision course with war against the Axis powers. World War One was the first war that regularly used submarine warfare. World War One also had the biggest sea battle in history, the Battle of Jutland (#1). The Allies fought German submarines with Q-ships, which were war ships disguised to look like merchant ships, only armed with massive ammunition in order to blow up submarines when they surfaced.

Yet Germany started hunting merchant ships aggressively in 1915, including American ships, and sank them with torpedoes. The German term "U-Boat" stood for "Unterseeboot." In fact, by February 4th, 1915, Germany had called for "unrestricted submarine warfare" against all Allied shipping in the British Islands. For example, in March 1915, the British ship *Falaba* was sunk by a German submarine, which killed 111 people (pg 361, #3). This act infuriated England, which wanted to implement a naval blockade of Germany. Then, on May 2nd, 1915, the passenger ship *Lusitania* left New York for England. A German submarine torpedoed the ship and killed 1,198 of the 1,959 people aboard (#1).

This act infuriated President Wilson, who saw the brutal German campaign of indiscriminately sinking passenger ships as unforgivable acts of terrorism (pg 87, #4). In response to Wilson, the Kaiser promised the President that Germany would sink no more ships. Unfortunately, this brutality continued when, on August 15th, 1915, the British passenger ship *Arabic* was sunk (#1).

Although many Americans were clamoring for war after these two incidents, Wilson would resist the temptation of entering the conflict in Europe, at least for the time being. Protests by Americans did prompt Germany to suspend submarine warfare for a time, but it did not last. Wilson's administration did attempt to negotiate its way out of war by making concessions to both the Germans and the English. England was disappointed with America's decision (pg 382, #3).

Around the same time, Wilson was also having problems with Mexico and its leader, Victoriano Huerta. Wilson refused to recognize Huerta's government, and waited for his government to be overthrown. The situation between Mexico and America grew tense when, in April 1914, Mexican authorities arrested U.S. sailors at Tampico unjustly. When U.S. officials demanded their release, they received their wish. Argentina, Brazil, and Chile participated in negotiating peace between Mexico and America. Huerta was forced to resign in July 1914, and was replaced by Venustiano Carranza, but the United States also did not recognize his government in 1915 (pg 423, #5). But Carranza also failed to take order of his country, and the revolutionary Pancho Villa controlled most of northern Mexico.

Villa crossed the U.S. border in March 1916, and he and his men attacked the town of Columbus, New Mexico, and killed 17 Americans (pg 423, #5). As a result, Wilson ordered General John Pershing and the 6,000 troops under his command to find Villa and his men and bring them to justice. Yet this mission was unsuccessful, and Wilson recalled the U.S. troops home by February 1917. Wilson's problems would only multiply during this time.

On March 25th, 1916, another English passenger ship, *Sussex*, was sunk by a German U-boat (pg 384, #3). However, Wilson was in a difficult position then regarding declaring war on Germany, because he campaigned for re-election in 1916 with the slogan, "He kept us out of war." This was even after the *Lusitania* attack (#1).

However, after Wilson's successful second presidential election in November 1916, the famous "Zimmermann note" was intercepted by British intelligence from German Foreign Minister Arthur Zimmermann in February 1917. In this letter, Zimmermann proposed to the Mexican government that an alliance should be formed between Germany and Mexico if the United States should enter the war on the Allied side. In return for this alliance, Zimmermann promised general financial support to Mexico, and to give Mexico the opportunity to reconquer the lost territory of New Mexico, Texas, and Arizona. This destroyed any public support for Germany in the United States and prompted Wilson to decide to enter the war (pg 423, #5). But in order to understand why President Wilson made this decision to go to war, knowing that lives and democracy itself were at stake, it is important to understand Wilson the person, and investigate the personal history that made him Commander-in-Chief. I will also investigate how Wilson became a successful war president against a formidable conglomerate, as well as how Wilson's ideological conflict with his political rivals failed him in the Treaty of Versailles and how these failures led to the rise of Nazi Germany and the subsequent war, 20 years later, that became World War Two.

President Wilson was born under the name Thomas Woodrow Wilson on December 28th, 1856. Wilson's heritage was Scotch-Irish, and he was a Presbyterian. Wilson had trouble in school as a child and was thought to suffer from dyslexia. Wilson was called Tommy as a child. He lived in Colombia, South Carolina in his youth. Despite Wilson's early grades, he was a diligent worker, and he learned to write shorthand notes in school, which helped him succeed.

Wilson's father, Joseph, was a Presbyterian minister, so he wished for the same profession for his son Woodrow when he came of age. But this clerical commitment was not meant to be. However, Wilson was a very religious Presbyterian, and he read the Bible every day, including throughout his later presidency (pg 412, #5). Wilson was thought to have learning disabilities as a child, including poor eyesight and trouble with arithmetic. Wilson did learn the basics of education from his parents.

Wilson began his education at Davidson College in North Carolina in 1873. Although he performed very well there, he was forced to withdraw after his freshman year, due to health reasons (pg 411, #5). When he regained his physical strength, Wilson's father resigned his post in North Carolina, and the family moved to New Jersey. Wilson enrolled at the College of New Jersey in 1875, which is now called Princeton University. While there, Wilson got involved in student government, and he became editor of the school newspaper, *The Princetonian* (pg 411, #5).

Wilson graduated from Princeton in 1879, achieving a rank of 38th out of 167 students. Wilson would enter the University of Virginia Law School in 1879. He was admitted to the bar in October 1882 (pg 412, #5). Wilson enrolled at Johns Hopkins University in Baltimore, Maryland in 1886, where he earned his Ph.D in political science. Wilson published his first book in 1885, which was called *Congressional Government*. This book was about Congressional control over the other two branches of government at the time.

Wilson met his first wife, Ellen, in April 1883, at his cousin's house in Georgia. They married in Savannah on June 24th, 1885 (pg 79, #3). During this time, Wilson criticized then-President Grover Cleveland as an ineffectual leader who was controlled by Congress. Wilson got a job out of his Ph.D program as a professor of political economy and public law at Bryn Mayr College in Pennsylvania.

In 1888, Wilson moved to Wesleyan College in Connecticut, where he remained until 1890. In 1890, Wilson became a faculty member at Princeton, where he taught jurisprudence and political economy. Wilson wrote a second book called *The State*, published in September 1890 (pg 101, #3). He was elevated to the presidency of the College of Princeton in 1902, the first layman in Princeton history to accomplish this feat (pg 414, #5). Wilson successfully reorganized university departments and academic curricula. Wilson was very popular in the Princeton community with both students and community leaders (pg 136, #3).

His intense popularity got Wilson nominated to the Governor of New Jersey Democratic ticket in 1910 on the first ballot (pg 414, #5). Wilson beat his Republican opponent, Vivian M. Lewis, on election night in November by a margin of 1,333,682 to 684,126 votes (pg 215, #3). As governor, Wilson passed the Geran Law, which mandated direct party primaries for all elected officials in the state and delegates to national conventions, and banned corporate contributions to political campaigns. Wilson also passed the state's first workers' compensation law (pg 415, #5).

Wilson was nominated for the 1912 Democratic presidential nomination at the Democratic convention, where he won on the 46th ballot (pg 415, #5). Wilson and his vice-presidential candidate, Thomas R. Marshall of Indiana, favored a lower tariff, enforcement of civil and criminal anti-trust laws, Constitutional protection of federal income tax, the banning of corporate contributions to campaigns, and recognition of the independence of the Philippines.

Wilson was running against incumbent President William Howard Taft and former President (and Bull Moose/Progressive candidate) Theodore Roosevelt. Wilson was helped by former President Roosevelt's entrance into the race of 1912, because it hurt President Taft more than it did Wilson (pg 258, #3). On election night, November 5th, 1912, Wilson won 42% of the popular vote (435 electoral votes), Roosevelt came in second with 27% of the popular vote (88 electoral votes), and Taft garnered 23% (8 electoral votes). This made Governor Wilson of New Jersey the 28th President of the United States.

Wilson named two-time presidential candidate and fellow Democrat William Jennings Bryan as his Secretary of State (pg 420, #5). Wilson legislated his "new freedom" economic programs through Congress, such as the Federal Reserve Act of 1913 and the Adamson Act of 1916, which established eight-hour working days and child-labor laws. Wilson created the Federal Trade Commission in 1914 to protect businesses from "unfair methods of competition" (pg 422, #5). This bureau was created to protect businesses and people from corruption in the United States. Wilson's motivation for his Clayton Antitrust Act was to legalize tools for unions to use for collective bargaining agreements, such as boycotts and strikes. It also barred corporations from acquiring stocks in another company. During Wilson's time as President, the 17th Amendment to the Constitution was passed in 1913, which provided direct elections of U.S. Senators by the people. The Eighteenth Amendment was the prohibition of alcohol, and the 19th Amendment extended the right of voting to women in 1920.

During this time, Wilson's Secretary of the Treasury, William McAdoo, married Wilson's daughter Eleanor on May 7th, 1914. Eleanor was 26 years younger than McAdoo (pg 322, #3). When U.S. soldiers were detained in Tampico in April 1914, this caused severe tension between the United States and Mexico. When the U.S. soldiers were released, Huerta consequently resigned. At the time, war was entirely possible between the United States and Mexico (pg 326, #3).

During this time, tragedy struck when Wilson's wife Ellen died of Bright's disease during his first term on August 6th, 1914. Wilson went into deep grieving over the loss of Ellen.

Yet by March 1915, Wilson had been introduced to Edith Bolling Galt, the widow of Norman Galt, a former Washington jeweler. President Wilson proposed marriage to Edith in May 1915, and they married at Edith's house on December 18th, 1915, while he was President. The U-boat sinking of the *Lusitania* and *Arabic* had taken place by then, and the Wilson Administration was attempting to make concessions to both the Germans and the English, much to the disappointment of England.

To make matters worse, German U-boats sank three more British ships. (the *Aboukin*, *Hogue*, and *Cressy*). Britain had lost a staggering 279,000 soldiers on the Western Front during 1915 (pg 101, #2). France had 1,292,000 casualties in 1915, and Germany had 612,000 casualties, but they still controlled northeastern France.

Pancho Villa led a raid in March 1916 over the New Mexico border town of Columbus that killed 17 Americans (pg 423, #5). In response, Wilson sent General Pershing and his 6,000 troops to capture Villa, but they were unsuccessful in bringing him to justice.

Wilson would campaign in 1916 for re-election as President under the theme, "He kept us out of war," which was a direct reference to the *Lusitania* tragedy. He also campaigned on military preparedness, a world association of countries to secure peace with nations and individuals, women's suffrage, pro-American unity, and retirement for federal employees. Wilson won a close second election over Republican nominee and Supreme Court Justice Charles Evans Hughes, 49% to 46% (277 electoral votes to 254 electoral votes).

While President Wilson pursued every peace negotiation possible to end the war in Europe, two events changed the course of history and catapulted the United States into fighting against the Axis powers in World War One. In February 1917, Germany announced that it would begin unrestricted submarine warfare against all shipping, commercial or otherwise. The second event was the interception of the Zimmermann note to Mexican political leaders, promising that if Mexico united with Germany to win the war, Mexico would re-acquire the U.S. lands of New Mexico, Texas, and Arizona. That would be the last straw for Wilson.

Wilson asked Congress for the passage of a bill to arm merchant ships in February 1917 in order to protect American commerce. Wilson would again appear before Congress on April 2nd, 1917, to ask for a Declaration of War against Germany, by declaring it "a war against mankind, and a war against all nations ... to fight for the ultimate peace of the world and for the liberation of its peoples, the German people included." Wilson made this statement at the end of his declaration because he stated that we were friends with the German people, but not their imperialist government (pg 92, #4). Wilson wanted to make the world safe for democracy.

By April 6th, 1917, both houses of Congress overwhelmingly passed the war resolution. The U.S. forces were called the American Expeditionary Force (AEF), and they were led by General John Pershing. Pershing previously served in the Spanish-American War at San Juan Hill and El Caney on July 1-3, 1898 (#1). Pershing was made brigadier general in September 1906 by President Theodore Roosevelt. Pershing's later victories in World War One enabled him to be promoted to a six-star general, the only soldier in American history ever to be so honored (#1). Pershing was given the order that he was to make military decisions independent of Allied control. An example of this was not allowing U.S. troops to be mercilessly slaughtered for 50 yards of territorial gain, as occurred in the Battle of the Somme.

Right at the beginning of America's participation in the war, the Allies set up mines in the ocean to thwart the submarines, and they were effective (pg 223, #2). Meanwhile, in Russia, Czar Nicholas the II's economy was in ruins, and as a result, Nicholas was forced from power in March 1917 (#1). After one unsuccessful coup, the Bolsheviks, led by Leon Trotsky and Vladimir Ilyich Lenin, led a successful Communist coup that made Russia a Communist state. These two men brokered a separate peace with Berlin and signed the Treaty of Brest-Litovsk, which ended Russia's participation in the war. In return, Russia yielded some territory to Germany, including Poland, Lithuania, Finland, Ukraine, and some Baltic provinces.

When Russia fell to the Bolsheviks in 1917, Wilson chose to ignore it, because the war against Germany was more important. Despite this, Wilson was certainly not happy about Russia's conversion, because that meant it was out of the war against Germany (pg 464, #4). Wilson never considered the United States a joint partner of Britain and France, and attempted to stay as independent as possible while assisting the Allies. This was very much unlike how President Franklin Roosevelt saw his situation during his war experience years later. Wilson observed that by the end of 1917, both England and France had suffered 5.8 million soldiers killed or wounded in this war, while Germany suffered 3,349,000 casualties (#1).

General Pershing believed in a more offensive war, with strictly American plans (#1). In between the dates of April 6th and June 26th, 1917, the first American troops arrived in France, and Pershing instituted specific individual training for his soldiers, in small-unit offensives. Meanwhile, on June 5th, 1917, the national draft registration was held in the United States, and more than 10 million men registered at 4,000 polling places across the country. The men ranged in age from 21 to 30.

Only 175,000 U.S. soldiers were in France by the end of 1917. General Pershing requested the U.S. government to raise an army of one million men for this war, and raised his request to three million men a few days later. The U.S. War Department began its first series of drafts on July 20th, 1917. President Wilson had entered the U.S. forces into the war because he felt that he had no choice, always first seeking a diplomatic solution over combat. However, as Wilson had officially committed troops to fight and die in this war, he felt that he had to convince the people that this war was a just cause, and that the soldiers' sacrifice would not be in vain.

By December 7th, 1917, America had also declared war on Austria-Hungary. On January 18th, 1918, Wilson proposed his Fourteen Points program of peace, which he wanted to implement when the war was won, to a joint session of Congress. Wilson hoped to, in the end, bring to the world a productive, peaceful, and just civilization (#1). The Fourteen Points mentioned previously will be listed at the end of the chapter.

During America's first offensive against the Central Powers, American troops helped win the Battle of Cantigny on May 28th, 1918. The American success continued in June/July 1918 in the Battles of Belleau Wood and Château-Thierry. That being said, however, the U.S. Marine losses at Belleau Wood were heavy.

Pershing's decision to train his troops for battle proved successful, despite European powers' desire to rush them off into battle. The AEF played an enormous role in the Second Battle of the Marne, which was also an Allied victory. In addition, 200,000 to 300,000 troops were sent to the war from America every month (#1). Because of the military victories of the First and Second U.S. Divisions, the German army withdrew from the Marne salient (#1). British and ANZAC forces (Australia-New Zealand Army Corps) beat back German forces on August 8th, 1918 (#1). This date would become known in Germany as "Black Sunday."

As a result, General Ludendorff ordered a withdrawal of the Lys salient and Amiens in northern France. This finally broke the stalemate on the Western Front in France (#1). This Allied Amiens offensive cost the German army 100,000 troops killed or wounded. Ferdinand Foch, who was placed in command of Allied troops on the Western Front, was responsible for defeating Ludendorff's 5th Offensive, because he turned this into an Allied counteroffensive that beat Germany out of the Marne salient (#1). The consequence of this defeat for Ludendorff was that he would order no more military offensives.

Next, the U.S. armed forces wanted to attack the German stronghold of St.-Mihiel, which was captured by the Germans in 1915 (#1). This U.S. offensive took place on September 12th, 1918, and it was an Allied victory; 15,000 German prisoners were captured here. The final U.S. offensive of the war was on September 26th, 1918, called the Meuse-Argonne Offensive. This battle was where U.S. soldiers saw the most combat during the war. This battle was in the Verdun area of France (#1). The Hindenburg line was the last line of defense for the Germans, which the Allies attacked. As early as October 6th, 1918, Germany and Austria-Hungary began sending notes to President Wilson that requested an armistice.

By November 7th, 1918, Austria-Hungary had surrendered. American and French military forces had already cleared the Argonnes Forest of Central Powers by November 3rd. By November 10th, the Kaiser had fled to the Netherlands. On November 11th, 1918, at the eleventh hour, Germany surrendered. On that day, Germany asked for peace on the basis of Wilson's Fourteen Points:

1.   Open covenants of peace, openly arrived at, after which there shall be no private international understandings of any kind, but diplomacy shall proceed always frankly and in public view.

2.   Absolute freedom of navigation upon the seas, outside territorial waters, alike in peace and in war, except as the seas may be closed in whole or in part by international action for the enforcement of international covenants.

3.   The removal, so far as possible, of all economic barriers, and the establishment of an equality of trade conditions among all the nations consenting to the peace and associating themselves for its maintenance.

4.   Adequate guarantees given and taken that national armaments shall be reduced to the lowest point consistent with domestic safety.

5.   A free, open-minded, and absolutely impartial adjustment of all colonial claims based upon a strict observance in principle that in determining all such questions of sovereignty, the interests of the populations concerned must have equal weight with the equitable claims of its government, whose title is to be determined.

6.   The evacuation of all Russian territory and such a settlement of all questions affecting Russia, as will secure the best and freest cooperation of the other nations of the world in obtaining for her an unhampered and unembarrassed opportunity for the independent determination of her own political development and national policy and assure her of a sincere welcome into the society of free nations under institutions of her own

choosing, and more than a welcome assistance also of every kind that she may need and may herself desire. The treaty accorded Russia by her sister nations in the months to come will be the acid test of their good will, of their comprehension of her needs as distinguished from their own interests, and of their intelligent and unselfish sympathy.

7.   Belgium, the whole world will agree, must be evacuated and restored without any attempt to limit the sovereignty that enjoys in common with all other free nations . No other single act will serve as this will serve to restore confidence among the nations in the laws which they have themselves set and determined for the government of their relations with one another. Without this healing act, the whole structure and validity of this international law is forever impaired.

8.   All French territory shall be freed and the invaded portions restored, and the wrong done to France by Prussia in 1871 in the matter of Alsace-Lorraine, which has unsettled the peace of the world for nearly 50 years, should be righted, in order that peace may once more be made secure in the interests of all.

9.   A readjustment of the frontiers of Italy should be effected along clearly recognizable lines of neutrality.

10. The peoples of Austria-Hungary, whose place among the nations we wish to see safeguarded and assured, should be accorded the freest opportunity of autonomous development.

11. Romania, Serbia, and Montenegro should be evacuated, occupied territories restored, Serbia accorded free and secure access to the sea, and the relations of several Balkan states to one another, determined by friendly council along historically established lines of allegiance and nationality and international guarantees of the political and territorial integrity of the several Balkan states, should be entered into.

12. The Turkish portions of the present Ottoman Empire should be assured a secure sovereignty, but the other nationalities, which are now under Turkish rule, should be assured an undoubted security of life and absolutely unmolested opportunity of autonomous development, and the Dardanelles should be permanently opened as a free passage to the ships and commerce of all nations under international guarantees.

13. An independent Polish state should be erected, which should include the territories inhabited by indisputably Polish populations, which should be assured a free and secure access to the sea, and whose political and economic independence and territorial integrity should be guaranteed by international covenant.

14. A general association of nations must be formed under specific covenants for the purpose of affording mutual guarantees of political independence and territorial integrity to great and small states alike.

World War One was an awful tragedy of carnage and mass violence, and one that President Wilson wanted the world never to repeat. Wilson himself arrived in France as head of the American peace delegation. The statistics of this grim conflict were astronomical for all sides involved:

1. France: 1,357,800 fatalities and 4,266,000 wounded
2. England: 908,371 fatalities and 2,090,212 wounded
3. Russia: 1,700,000 fatalities and 4,950,000 wounded. These statistics are even worse when considering two million civilian deaths during this same time period.
4. Italy: 462,391 fatalities and 953,886 wounded
5. Germany: 1,808,546 fatalities and 4,247,143 wounded
6. Austria-Hungary: 922,500 fatalities and 3,620,000 wounded
7. United States: 50,585 fatalities and 205,690 wounded
8. Turkey: 2,150,000 civilian deaths, which included victims of the Armenian massacre (#1)

The Paris Peace Conference began without representatives from the Central Powers. Delegates present from the victorious nations at the conference formally approved Wilson's request that the League of Nations be made an integral part of the peace treaty. America was the clear victor, as well as the decisive factor, in the war, and held power over other European nations (pg 22, #7). Wilson wanted to form the League of Nations in order, specifically, to prevent future wars (pg 531, #3).

France changed leaders in November 1917 when Georges Clemenceau became the Premier of France (#1). Clemenceau acknowledged that without both England's and America's help, France would have lost the war (pg 32, #7). England was represented by their Prime Minister, David Lloyd George. There was a sense of massive resentment against Germany by the Allied powers during the Paris Peace Conference. These three nations met often from January to June 1919 (pg 37, #7). A fourth leader joined them in Paris: Premier Vittorio Orlando of Italy.

The negotiations were so intense that President Wilson actually suffered a minor stroke in April 1919 (pg 556, #3). The official Treaty of Versailles was signed by all parties on June 28th, 1919, which was the five-year anniversary of the assassinations of Archduke Ferdinand and his wife, Sophie. This treaty put all responsibility for the war on Germany, and stripped it of its colonial possessions, including the Saar Basin and the Alsace-Lorraine region, which was awarded to France. Also, the Sudetenland was given to Czechoslovakia, some territory given to Belgium, and the Polish corridor to Poland, which was further established by the Treaty of Riga in 1921.

In fact, the Treaty of Riga, signed on March 18th, 1921, gave Poland a border in the east after a failed Bolshevik invasion of Warsaw, and Poland added 4 million Ukrainians, 2 million Jews, and 1 million Belorussians. In 1919, the Treaty of Neuilly-sur-Seine gave western Thrace to Greece from Bulgaria. On April 16, 1922, the Treaty of Rapallo, signed by Germany and Russia, set up the borders between Italy and Yugoslavia. In 1919, the Treaty of St.-Germain-en-Laye gave Austria a strip of territory from the western edge of Hungary. Austria was the only defeated nation to actually gain territory. The Treaty of Sèvres in 1920 promised independence to Armenia from Turkey. In 1921, the Treaty of Moscow established the borders of Russia and Turkey. Finally, in 1923, the Treaty of Lausanne established Turkish borders from eastern Thrace to Syria.

As far as dealing with Germany, the Rhineland was demilitarized and the German armed forces were to constitute no more than 100,000 troops (#1). Essentially, the three rules of treatment of Germany after World War One were:

1. punishment
2. payment
3. prevention of future aggression

The meeting in Versailles by the Allied powers did agree to put the Kaiser on trial (pg 558, #3). They also agreed that Germany should financially repay the Allies $35 billion for causing World War One, to be paid out over a 30-year period. As history would record, the Treaty of Versailles punished Germany so hard that it created the desperate economic and social conditions that would give Hitler his chance to rise to power. Yet power was initially given to Socialist Gustav Adolf Bauer immediately after the war.

President Wilson would face even bigger problems at home, when the Republican party, led by Senator Henry Cabot Lodge of Massachusetts, took control of both houses of Congress. This made Wilson's job of passing his legislation through Congress much more difficult, including the Treaty of Versailles and the League of Nations, both of which Congress rejected. While President Wilson embarked on a national speaking tour through the country to win support for his legislation, he suffered a major stroke on September 25th, 1919 in Pueblo, Colorado. Wilson would never achieve full physical recovery from this.

Because the Constitution does not provide guidance in coping with a disability affecting a sitting president, the United States faced a crisis. Though Wilson did consider resigning, he eventually chose not to do so. Neither would Vice President Thomas Marshall, or Congress would challenge Wilson for this outcome. Despite the fact that the United States would never join the League of Nations, Wilson was awarded the Nobel Peace Prize in 1919 (pg 425, #5).

Wilson's wife Edith brought presidential work to his sickbed, and Edith decided which matters were important enough to bring to the President, and which were not (pg 425, #5). Two amendments to the Constitution were passed, and they were the 18th Amendment, which prohibited alcoholic beverages, and the 19th Amendment, which gave women the right to vote. Yet Wilson continued to fight Congress by vetoing its resolution to an end of hostilities with Germany and Austria-Hungary in May 1920.

In July 1920, the Democrats nominated James Cox of Ohio for President, and Franklin Roosevelt for Vice President. Together, they campaigned on a platform that called for the 1920 election to be a "solemn referendum" on the Treaty and the League. Yet they would be defeated on November 2nd, 1920, by Republican Warren Harding. Under Harding's administration, he would sign a Congressional Joint Resolution that declared an end to the war with Germany. Also, the Treaty of Trianon was signed with Hungary in June 1920 (pg 316, #2). At the time, Britain controlled Palestine (pg 320, #2). Wilson's vision of a League of Nations did later influence the United Nations that was established immediately following World War Two.

Wilson would only live until February 3rd, 1924, suffering from illness (pg 673, #3). To Wilson's credit, the United States did emerge from World War One as a world power. The Allies would not have won World War One without America's military participation. Despite this enormous victory by the United States, it would only take 20 years and 67 days from the official end of war, on June 28th, 1919, until England and France would be at war with a new, and more evil, leader of Germany, who would unleash the greatest conflict and carnage the world has ever seen (pg 323, #2).

Bibliography

1. *The War to End All Wars: The American Military Experience,* by Edward M. Coffman (University of Wisconsin Press, 1986).
2. *A Short History of World War One*, by James Stokesbury (William Morrow, 1981), disc 3.
3. *Woodrow Wilson: A Biography*, by August Heckscher (Macmillian, 1993), disc 5.
4. *World War One: Opposing Viewpoints*, by William Dudley (Greenhaven Press, 1998), disc 3.
5. *The Complete Book of U.S. Presidents*, by William A. Degregario.
6. *The Wars of America*, by Robert Leckie.
7. *Paris 1919: Six Months That Changed the World*, by Margaret MacMillian (Random House, 2001), disc 3.
8. *Woodrow Wilson*, by H.W. Brands (Times Books, 2003).
9. *Almanac of World War I*, by David F. Burg and L. Edward Purcell (University Press of Kentucky, 1988)
10. *The American Way of War: A History of United States Military Strategy and Policy*, by Russell F. Weigley (Indiana University Press, 1977).
11. *The Military History of World War I*, by Trevor N. Dupuy (Watts, 1967).

Chapter 10
The Study of World War Two and the Actions and Decisions of Both Presidents Franklin Roosevelt and Harry Truman (December 7th, 1941-August 15th, 1945)

The most horrible and catastrophic event in world history, known as World War Two, which occurred from 1938 to 1945, is argued by some historians as the continuation of the same conflict fought 20 years earlier during World War One (#1). The United States was directly involved in this war from December 7th, 1941 until August 15th, 1945. The difference was that the Second World War surpassed the horrors and carnage of the First World War, involving the military participation of more than 60 countries and the murders of at least 57 million people globally, more than half of whom were civilians. World War Two is described as "the most destructive human endeavor in history" (#1). It is also described as the most significant event in world history, and one that continues to affect our lives to this very day.

Though this war did not end all wars after 1945, it did bring the United States out of its economic depression and into a period of growth. America was also successful in providing some global stability until the end of the century (#1). This war featured the Allied Powers (England, Russia, and the United States) versus the Axis Powers (Germany, Italy, and Japan). This is ironic, because both Italy and Japan had fought against Germany and Austria/Hungary during World War One. In fact, Japan had entered World War One against Germany because Germany had refused to cede territory it had claimed in northeast China (#1).

Yet when World War One was over, by November 11th, 1918, President Woodrow Wilson received no cooperation in passing his League of Nations legislation from the Republican majority in the Senate. This group included Hiram Johnson of California, Henry Cabot Lodge of Massachusetts, and William Borah of Idaho, who fought Wilson vigorously on his legislation (#1). As a result, not only did the United States not join the League of Nations, but it also failed to sign the Treaty of Versailles.

140

Historians argue that had the United States joined the League of Nations at the time, it could have, in theory, prevented the course of events that gave birth to the rise of the evil Nazi regime in Germany. However, the United States did not participate freely in the world community, due to the isolationist beliefs in the country at the time, which focused on America's domestic struggles first. Nor did the United States support its European allies in their effort to contain Germany, which led directly to the rise of the Nazis. This left both France and England as the dominant powers in the League of Nations.

Yet they each had different ideas about how to deal with Germany after World War One. England wanted to include Germany in the world community, but France wanted to keep Germany isolated and punished. As a result, France signed a defensive alliance with Poland in 1921. This was followed by a French alliance with Czechoslovakia in 1924, Romania in 1926, and Yugoslavia in 1927. Meanwhile, Germany signed an economic agreement with Russia in 1922, in an effort to improve both nations (#1).

The reason why Lenin, and later Stalin, pulled out of World War One when they did was because they wanted to rebuild the Imperial Army into a newer, stronger, Red Army that could perform much better in a later war (pg 21, #1). America returned to isolationism in its foreign policy during the 1920s, and concentrated on domestic affairs (#1).

Even after its defeat in World War One, Germany still remained a power in Europe (#1). However, the German economy had collapsed by 1923. In this climate of political ambiguity, a man named Adolf Hitler would enter the scene in German politics, and his lust for power would soon be fulfilled by several tragic circumstances. Hitler was born on April 20th, 1889, in Braunau am Inn, Austria (#1). As a youth, Hitler showed interest in being an artist, but he was rejected twice in applying to Vienna's Academy of Art.

Despite the fact that Hitler was an Austrian citizen, he fought with the German Army during World War One. He only made it to Private First Class (#1). In 1923, Hitler seized power by overtaking a beer hall with his politicians, who were there for a meeting. The word *Nazi* comes from the word *Nasos*, which was the term "National Socialist" abbreviated (#1). Hitler was captured with his followers at the beer hall, and was sentenced for his crime to five years in Landsberg Fortress. He wrote the book *Mein Kampf (My Struggle)* while he was in prison. While Hitler was incarcerated, he capitalized on the misery of the public when the German economy collapsed (#1).

The Nazis actually won seats democratically in Germany during 1928 and 1930. Hitler himself did not even become a German citizen until February 1932, when he ran for president (#1). Although Hitler lost that election, he had a lot of support from Germans, and he also had his own private army of 400,000. President Paul von Hindenburg gave Hitler the role of "Chancellor" on January 30th, 1933 in the hopes of containing and appeasing the Nazis. Unfortunately, Hitler took power of Germany early in 1933 and began his wicked dictatorship of the country immediately.

Once in power, Hitler outlawed the Communist Party and trade unions and murdered all of his political opponents (#1). It took Hitler only a few months to separate Germany from democracy. When Hindenburg died in 1934, Hitler became Germany's President. Hitler was then the "Führer," or leader. Hitler said that the main enemy was the Jew, all of whom should, therefore, be eliminated.

Hitler began his persecution of the Jews by boycotting their businesses, then stripped them of their jobs, their property, and their rights (#1). The Gestapo (secret police) had the power to arrest anyone. Hitler opened the first concentration camp, at Dachau, in 1933. Thankfully, Hitler had no children (#1). Meanwhile, America's leader in 1933 was the newly elected Franklin D. Roosevelt.

Roosevelt was born on January 30th, 1882, at his family home in Hyde Park, New York (pg 482, #4). Roosevelt attended the prestigious Groton school from 1896 to 1900 for high school, and his cousin Teddy, who was then Governor of New York, spoke at his graduation (pg 483, #4). Franklin Roosevelt would go on to attend Harvard from 1900 to 1904. He majored in political history and government, but he was also involved in the glee club and the Harvard Union Library Committee, and, most important of all, he became editor-in-chief of the Harvard newspaper, the *Crimson* (pg 483, #4).

Roosevelt actually had the credits to graduate after only three years, but he chose to stay a fourth year to continue his work for the *Crimson*. This experience served him well in later years. After Harvard, Roosevelt went to Columbia Law School from 1904 to 1907 (pg 483, #4). Roosevelt cast his first ballot for president in 1904, for his cousin Theodore, who was then the sitting president (pg 25, #3). His wife, Eleanor, was the daughter of Teddy's brother Elliott and his wife, Anna Hall. In fact, Eleanor was Franklin's fifth cousin. Franklin and Eleanor were married on March 17th, 1905 (pg 26, #3). President Theodore Roosevelt gave the bride away at the ceremony.

Their first child, Anna, was born in May 1906 (pg 27, #3). Franklin became involved in politics in the 1910 elections, in which he won a state senate seat from New York (pg 34, #3). That election saw huge numbers of Democrats win offices during President Taft's administration. Because Franklin was a Democrat, he supported Woodrow Wilson for president in 1912, even against his own cousin Theodore, from the Bull Moose/Progressive Party. In fact, Theodore was shot during the 1912 presidential campaign by John F. Schrank in Milwaukee, who was an anti–third-term fanatic. Nevertheless, Roosevelt would go on to give his speech anyway for over an hour, with the bullet lodged in his chest (pg 388, #11).

Theodore Roosevelt finished second in the 1912 presidential election, behind Woodrow Wilson but also ahead of William Taft, the incumbent president. President-elect Wilson named Franklin as Assistant Secretary of the Navy in 1913 (pg 50, #3). At this time, Franklin's wife, Eleanor, hired 22-year-old Lucy Page Mercer as her social secretary. It is unknown exactly when Franklin and Mercer's love affair began, but Eleanor caught the two exchanging love letters in 1918, which almost led to their divorce.

Franklin did agree to end his relationship with Mercer after being confronted by Eleanor, who gave him the ultimatum of divorce or reconciliation. Eleanor then assumed a more independent role from her husband. In 1920, Mercer would go on to marry a wealthy widower named Winthrop Rutherfurd. Yet she and Franklin continued their affair all the way into his presidency, until Franklin's death in 1945 (pg 485, #4).

When Franklin was appointed Assistant Secretary of the Navy, in 1913, he found dealing with unions to be a major challenge. Nevertheless, he worked well with them. As a result, there were no labor strikes during Roosevelt's years as Assistant Secretary of the Navy (pg 53, #3). Roosevelt recommended to President Wilson that the United States become involved in World War One as early as the fall of 1916 (pg 61, #3).

After Roosevelt's success as Assistant Secretary of the Navy, he was awarded with the vice-presidential nomination on the Democratic ticket in 1920, with James Cox as the presidential candidate. They campaigned under the League of Nations approval platform, which had such difficulty gaining passage in both houses of Congress during the Wilson Administration. They were defeated in the 1920 presidential election and lost to Warren Harding.

A much bigger tragedy happened in Franklin's personal life in August 1921, when he went for a swim in the Bay of Fundy, which contaminated him with the disease of poliomyelitis, crippling him for the rest of his life (pg 87, #3). This experience changed Franklin's character profoundly (pg 89, #3). Though he was not involved in politics from 1920 to 1928, FDR was involved in a law practice with his law partner Basil O'Connor in New York City (pg 486, #4). Because of Franklin's polio affliction, his wife Eleanor became more involved in his political life (pg 91, #3).

In 1928, Franklin decided to run as a Democrat for the governorship New York against State Attorney General Albert Ottinger. Franklin won this election, although it was by a narrow margin (pg 486, #3). Franklin was successful as governor in protecting workers' rights by limiting the workweek to no more than 48 hours and extending workers' compensation (pg 486, #3).

Because of Roosevelt's success as the Governor of New York, he decided to run for the presidency as a Democrat in 1932. He would be going against incumbent President Herbert Hoover. It was at the Democratic nomination acceptance speech that Roosevelt introduced his

"New Deal" initiatives to bring America out of its economic depression. During this time, the Great Depression was in full swing under President Hoover's administration.

As a result, on November 8th, 1932, Franklin Roosevelt beat incumbent President Herbert Hoover by a popular vote of 57% to 40%, and an electoral college tally of 472 to 59 (pg 488, #4). The New Deal program was Roosevelt's war against the depression (pg 18, #2). Another potential problem for Roosevelt when he became president was that the U.S. Army had also felt the effects of the Great Depression. Consequently, the United States had only the 18th most powerful army in the world, behind such powers as Germany, Britain, France, Russia, Italy, Japan, China, Spain, Sweden, Switzerland, etc. (pg 23, #2). However, the United States did lead the world in the production of automobiles, washing machines, and other household items.

Along with Roosevelt's successful presidential victory in 1932, the Democratic Party also won back control of both the House and the Senate (pg 144, #3). Franklin once said that the office of the presidency is "a superb opportunity for applying, and reapplying in new conditions, the simple rules of human conduct to which we always go back. Without leadership alert and sensitive to change, we are all bogged up or we lose our way" (pg 151, #3).

Roosevelt's first task as president was to fix the economic depression that plagued America. In fact, Roosevelt got leaders from both political parties to form a national partnership to promote economic recovery (pg 197, #3). Roosevelt was responsible for overturning the alcohol ban (Prohibition) with the 21st Amendment (pg 168, #3). On the other side of the world, at the same time, Hitler had destroyed German democracy in less than one month. Yet even Hitler himself praised Roosevelt's economic plan for America in 1933 (pg 184, #3). Also at this time, the Japanese military began its military aggression against its Asian neighbors. A political advantage opened for President Roosevelt when the Democrats increased their numbers in both houses of Congress in the 1934 elections (pg 202, #3).

Nevertheless, Roosevelt had trouble passing his New Deal legislation because the conservative majority of the Supreme Court struck down the laws with their decisions. This made Roosevelt aware that he would have to fight the Court's decisions for the rest of his presidency. Yet as far as Roosevelt's bipartisanship is concerned, it is said of him, "He was a leader of all people, and he was perfectly willing to subordinate the interests at the same time that all other interests found an equal place in the national plan" (pg 237, #3).

Another quandary for Roosevelt was that isolationism was strong in America, especially in the American Midwest. Consequently, the isolationists wanted Roosevelt to stay neutral in the European military conflict, so he did. Roosevelt's New Deal legislation was responsible for creating 6 million new jobs in 1936 (pg 266, #3). President Roosevelt would go on to beat Republican candidate Alf Landon in the 1936 presidential election handily (pg 490, #4).

Unfortunately for the re-elected Roosevelt, the crisis in Europe was growing to a fever pitch. In 1936, Hitler wanted to make Germany the most powerful army on Earth, and his dictatorship policies set out to do just that. Hitler broke the Treaty of Versailles stipulation that there would be no German army (#1). Unfortunately, people around the world, including Americans, did not take the impending threat of Hitler seriously enough until it was too late.

Consequently, the Nazis retook the Rhineland that Germany had lost after World War One on March 7th, 1936 (#1). The U.S. military did not stop Hitler then because of the country's economic and social problems and because the U.S. military had been significantly downsized at the end of World War One. Hitler would capitalize on this fact (#1). Also by January 1938, there were still eight to eleven million people who were unemployed in America. Roosevelt insisted on creating a balanced budget by the year 1940 (pg 324, #3).

Even the threat posed by the Nazis in Europe was not Roosevelt's biggest problem. By May 1936, Italy had invaded Ethiopia, though it took five years to take complete control of that country. Japan also proved to be an international and aggressive threat to democracy by invading Manchuria in 1931 (#1). Japan was the only country besides the United States to emerge stronger after the First World War, and the Japanese were eager to expand their Asian empire. While China complained of Japan's aggression to the League of Nations, the League was powerless to prevent Japan's actions.

By July 1937, Japan had invaded China and conquered the most lucrative parts of the country. Japan also issued threats to the Western nations by warning against helping China in any way while it was being conquered (#1). Consequently, the Japanese aggression against American naval vessels began as early as December 1937, when Japan bombed and sank the U.S. gunboat *Panay* in Chinese waters. This act killed two U.S. sailors and wounded thirty. When this incident occurred, the Japanese government apologized to President Roosevelt, so the United States did not respond militarily against Japan at that time. This inaction by the more isolationist U.S. government may have sown the seeds for the later Pearl Harbor attack on December 7th, 1941.

Japan continued its aggression unopposed, invading and conquering the city of Nanking. This act by the Japanese military was one of the most awful, brutal, and horrible acts by humanity in recorded history, in which soldiers raped 20,000 Chinese women and murdered 300,000 innocent people. This was one of history's worst massacres (#1).

At the same time, Nazi aggression towards its Austrian neighbor began with the assassination of Chancellor Engelbert Dollfuss. The Nazi killers were arrested, and a new Chancellor, Kurt von Schuschnigg, took over control (#1). But once the Italian leader Benito Mussolini joined Hitler's alliance, Austria was vulnerable to be conquered by Germany, which occurred on March 12th, 1938. As a result, Hitler promised Mussolini his full allegiance (#1). Once Germany conquered Austria, the newly acquired territory was renamed Ostmark (Eastern March) (#1). The treaty that Germany and Italy formed was a direct violation of the Treaty of Versailles, but would continue to go unpunished.

Immediately, Jewish people were mercilessly persecuted, and their businesses were closed: 75% of the country's Jewish population fled the country (approximately 185,000 people). Among those fleeing was Sigmund Freud. However, four of his sisters did not leave, and were consequently murdered in concentration camps (#1).

Next on Hitler's list for conquest was the country of Czechoslovakia. The Sudetenland, which was a German-populated part of Czechoslovakia, was ceded to Germany. This act was done in fear of Germany showing more military aggression, thereby making more territorial demands. The Treaty of Munich was supposed to satisfy Hitler's greed for land acquisition, but he was untruthful and had full intention of taking all of Czechoslovakia (#1). Hitler then conquered the entire country of Czechoslovakia by March 16th, 1939 (#1).

Hitler's next target was Poland, with which he had retracted the 1934 non-aggression treaty. This alerted both France and England that they would have to align themselves with Russia if Germany attacked Poland, which they eventually did. After all, it was Germany that went through Poland to attack Russia during the First World War. The Soviet Premier at that time was a man named Joseph Vissarionovich Djugashvili, otherwise known as Joseph Stalin. Stalin, in the Russian language, means "man of steel" (#1). Before the Bolshevik Revolution of 1917, Stalin was exiled by the Czarist government for supporting the Communists, but he was brought back from exile when Lenin took power.

Stalin succeeded Lenin in 1924, after he outmaneuvered his political rivals. Sensing the inevitable German invasion of Poland, Hitler and Stalin signed a pact that allowed Hitler to attack Poland, while agreeing to leave Russia alone, on August 23rd, 1939 (pg 684, #10). In return, Russia had permission from Hitler to invade Finland, because the country would provide vital resources to Germany (#1). Consequently, the war in Poland began when German soldiers, dressed like Polish soldiers, attacked a German radio station. This would give Hitler all the reason he needed to launch his blitzkrieg, which marked the official beginning of World War Two on August 31st, 1939 (#1).

Terrorism would rear its ugly head on September 3rd, 1939, when a German U-boat sank the British passenger ship *Athenia*, which killed over 1,200 people, including 28 Americans. This terror act was followed by 12 more sinkings (pg 694, #10). Poland was soon overrun, but Hitler had betrayed his promise to Stalin. As bad as the situation in Europe appeared, America had decided to remain neutral, and stayed on the sideline for two more years (#1).

Joseph Kennedy was the U.S. ambassador to England in those days, and he reported to President Roosevelt that the situation in Europe was serious. Despite countries in Europe falling rapidly into the hands of Germany, Ambassador Kennedy advised Roosevelt to remain neutral (pg 26, #2). Though Roosevelt was initially opposed to the war, he saw that the German military forces were superior to those of the rest of Europe. By September 3rd, 1939, England had declared war on Germany. Winston Churchill was the Prime Minister of England by 1940. He had been in the English Parliament since the year 1900 (pg 33, #2). Churchill and Roosevelt were also good friends and strong allies.

Churchill suggested to Roosevelt that the United States should begin to build up weapons to prepare for an attack from Hitler in his lust for conquest. Roosevelt wanted to send aid to the Allies, but stopped short of joining the war effort. Roosevelt did, however, want the existing Neutrality laws changed. As a result, Roosevelt decided that the United States would sell war items by cash, so that European debt to the country could be avoided, also greatly benefiting the U.S. economy (#1). After all, in 1940, the depression was technically in its eleventh year of existence, and Roosevelt envisioned fixing the U.S. economy by building military armaments to help the Allies fight the Germans (pg 46, #2).

At the same time, the money the United States acquired from the sales of weapons to the Allies was used to build up the strength of the U.S. Army, moving it up considerably in the world rankings. This was necessary because, at the time, Germany had better tanks than America did.

After Hitler's conquest of Poland, his next targets of invasion were Norway and Denmark (#1). Hitler conquered Denmark on April 9th, 1940, after no fighting from the Danes (#1). Norway attempted to fight the invading Nazis, but they, too, were conquered (#1). Hitler then attacked Belgium and the Netherlands on May 10th, 1940, with 136 German divisions. Ironically, Churchill became Prime Minister of England on that very same day (#1). It would take Hitler four days to conquer the Netherlands, from May 10th to the 14th, 1940 (pg 82, #1). Hitler then took Antwerp in Belgium on May 18th, 1940. Next was Hitler's most important prize, the country that had punished Germany so vehemently after World War One: France. When Germany invaded and conquered France in May and June 1940, Italy invaded France also (#1).

Hitler was obsessed with punishing France the same way France punished Germany at the end of World War One. The battle in France began at the Battle of Dunkirk in northern France, which left 60,000 British soldiers dead, wounded, or captured (pg 62, #2). Dunkirk was completely overrun by Germany. Consequently, France surrendered to

Germany completely on June 14th, 1940. Yet Germany would only take the northern part of France, letting Italy occupy the southern half. Hitler further charged France $120 million for damages sustained to Germany.

General Charles de Gaulle fled to England, and Prime Minister Churchill evacuated 340,000 English troops from Dunkirk, which helped England fight off a Nazi invasion. Churchill felt that he had to protect himself and the English people from imminent German aggression and invasion. Historically speaking, it took Hitler less than one year to conquer most of Europe (#1). Churchill pledged to his citizens, and to the world, that he would never surrender to Hitler and the Nazis, no matter what. Yet in order to successfully achieve this, he knew that he needed help from the United States (pg 705, #10).

To counter the threat posed from Europe, Roosevelt needed to initiate a military service draft, and federal money had to be appropriated for defense (pg 711, #10). Therefore, Roosevelt officially asked Congress for both in September 1940 (#1). Interestingly enough, Roosevelt was also fighting another political battle at home, this one involving his running for an unprecedented third term as president. If successful, he would be the first American executive in history to do so.

When the Democrats met at their presidential convention in July 1940 in Chicago, the delegates were confused as to whether Roosevelt would, in fact, seek a third term (pg 490, #4). Postmaster General James Farley and Vice President John Garner also were running for the nomination, believing that Roosevelt would retire from the presidency after two terms (pg 490, #4). Instead, Roosevelt allowed his name to be re-nominated at the convention, issuing a statement through Senator Alben Barkley of Kentucky that though he was not necessarily seeking a third term, he would accept it if the American people wanted him to (pg 490, #4).

Roosevelt would consequently win his party's nomination for a third term as president on the first ballot (pg 490, #4). He did change his choice for vice president, however, by replacing John Garner with Agriculture Secretary Henry Wallace of Iowa. Roosevelt's platform consisted of his New Deal accomplishments and the pledge to keep the United States out of the war in Europe. This is very similar to the campaign run by then-President Wilson 24 years earlier. But Roosevelt sensed the accumulating aggression from Europe, and he consequently vowed to build and maintain a strong defense as a deterrent to the Axis powers. His Republican opponent for president was Wendell Willkie, who was once a Democrat as recently as the elections of 1938. Willkie campaigned on limiting the presidency to two terms, the failure of the New Deal policies, equal rights for women, and taking a stronger stance against Hitler (pg 491, #4).

The New Deal was criticized by business leaders for supporting welfare states and organized labor (pg 53, #2). Roosevelt countered this statement by attempting to reach out to business leaders of both parties by involving them in Cabinet positions and creating materials (pg 54, #2). For example, in order to successfully run for president in 1940, Roosevelt named two Republicans to his Cabinet: Henry L. Stimson as the Secretary of War and Frank Knox as the Secretary of the Navy, even though Knox was the running mate of his 1936 Republican opponent, Alf Landon (pg 424, #3). Consequently, Roosevelt successfully merged the business world with government (pg 57, #2).

Roosevelt would also use "fireside chats" on the radio to talk to the American people directly to explain himself. Despite this, some business leaders did not like the New Deal anyway, and the war was not looking good in Europe. Nevertheless, on election night, November 5th, 1940, Roosevelt defeated his Republican opponent, Wendell Willkie, by a vote of 55% to 45%, or 449 electoral votes to 82 (pg 492, #4). While Roosevelt celebrated his unprecedented third consecutive term as president, he got right back to work in cooperating with British Prime Minister Winston Churchill in helping Britain build up military protection against an assumed invasion from Germany. This protection was sought because England was practically the only European power left by then. Ironically, Hitler never conquered Sweden or Switzerland (#1). Churchill asked the United States for unconditional military support to help fight Germany, and Roosevelt completely agreed (pg 64, #2). Roosevelt sent Churchill war material such as bomber planes, rifles, tanks, machine guns, and rounds of ammunition.

Meanwhile, Hitler was showing off in the newly conquered France by marching through all of the historical sites of Paris with Nazi flags flying through the streets. Then, the Nazis persecuted the Jewish French people and forced them to go to concentration camps. While the Jews were being persecuted by the Nazis, many fled the conquered territories for other countries. The United States took in 105,000 refugees, but no more; this would turn out to be a controversial decision by Roosevelt. Palestine took in 55,000 refugees. Tragically, this escape did not happen for six million Jewish people in Europe, who were mercilessly killed by the evil regime of Hitler.

The Nazis were also thought to have stolen $2.5 billion worth of art by the end of their terror regime in 1945 (#1). After Hitler's successful conquering of France, he then turned his aggression towards England. Hitler began bombing the country with Luftwaffe airships by July 1940. This attack was called, in code, "Operation Sea Lion" (#1). Roosevelt proceeded with his implementation of sending massive armaments to England to assist in fighting off the German invasion.

Roosevelt had the draft instituted on October 16th, 1940, to help construct an American military to respond against possible foreign aggression. Consequently, 16 million Americans signed up. Yet Roosevelt insisted that the United States would stay out of the fighting (#1). Roosevelt envisioned that by arming European allies against Hitler, the

United States was fighting aggression and for democracy (#1). By lending destroyers and other armaments to England, Roosevelt helped England successfully thwart Hitler's inevitable plan to invade it by crossing the English Channel (pg 441, #3). In return for giving Britain fifty destroyers for battle, England gave the U.S. military the use of nine English bases, which proved to be greatly beneficial.

When confronting domestic policies, Roosevelt gave tax breaks to businesses that helped with manufacturing supplies for the army, much to the dismay of liberals. Also at home, First Lady Eleanor Roosevelt led a coalition to give African-Americans more jobs and opportunities in society. Eleanor was heavily in favor of anti-lynching laws (pg 163, #2). The African-American community had backed Franklin for president in 1936 and 1940, despite living in a very racist society where they were denied fair opportunities in life and job growth (pg 165, #2). For example, black people could not apply for the Army Air Corps. Yet while First Lady Eleanor was trying to help African-American citizens at home, Franklin was more reluctant in pursuit of this cause (pg 170, #2), but he did promote Ben Davis from colonel to brigadier general and bring Dean William Hastie of Howard Law School to the War Department (pg 172, #2).

Due to the U.S. development of weapons to assist the Allied war effort during this time, the U.S. economy had significantly improved. It was important that Roosevelt helped Britain with massive financial assistance, because the United Kingdom was almost in a state of bankruptcy. Despite this, Roosevelt was completely loyal to England and his friend Winston Churchill. The U.S. population backed Roosevelt's policies by 88% to send money to England to fight Germany, but did not want to get directly involved in the war campaign. But in Europe, Hitler's aggression was getting worse and ever closer to England's borders. Because Germany had no real navy to speak of, an amphibious assault on the country of England would be very difficult. So, instead, the Nazis started a blitzkrieg bombing on English cities and civilians. German night bombing of England was done with the intention of being more terrifying, but it proved to be inaccurate at hitting targets. In response, England attacked Berlin (#1).

Sensing the inevitable threat that Hitler posed to freedom, on January 6th, 1941, Roosevelt issued a proclamation to the world for all human beings to have the following freedoms: freedom of speech, of worship, from want, and from fear (pg 201, #2). But this would not prevent Hitler's relentless ransacking of European countries. In April 1941, Germany invaded Yugoslavia, killing 17,000 civilians within 24 hours (pg 233, #2). Yugoslavia would capitulate four days later.

Next, Hitler invaded Greece, and he conquered the entire country in less than four weeks (pg 233, #2). German submarines were sinking Allied boats three times as fast as they could be built. On May 27th, 1941, Roosevelt announced the immediate strengthening of American defenses against Hitler and Mussolini. On the same day, Britain torpedoed and sank the German ship *Bismarck*, killing 2,300 German sailors (pg 107, #1). On June 4th, 1941, Roosevelt advisor Missy LeHand had a physical breakdown and fell ill (pg 244, #2). Roosevelt often referred to her for advice on presidential matters. She opened the president's views on civil rights, as did First Lady Eleanor, by witnessing the African-Americans march in Washington. This made a profound impression on Roosevelt, and it made him want to curb the economic enslavement of African-Americans and give them more opportunity in society.

Meanwhile, back in Europe, events would take a turn for the worse between German and Russian relations when Germany invaded Russia on June 22nd, 1941. Hitler put 150 divisions on the Russian border, which equaled three million men. Hitler also installed 2,700 planes, 3,300 tanks, and 600 motor vehicles. There would be hundreds of thousands of Russians killed in this German assault. During this time, Hitler convinced Romania and Bulgaria to join the Axis (#1), while Britain occupied Iran in 1941.

In September 1941, the Nazis went on to capture Kiev in Russia and 660,000 Russian prisoners (#1). Hitler ultimately failed in his invasion of Russia, similar to Napoleon's failure doing the same thing 130 years earlier (#1). Roosevelt secured federal aid to help Stalin's Russia by sending it massive armaments and supplies in a land-lease act conceived by the president.

While all of this was going on, the empire of Japan was certainly becoming dogmatic in the global community, especially in Asia. By mid-July 1941, 40,000 Japanese troops invaded the rubber-rich territory of Indochina. As Roosevelt was suspicious of Japanese motives, he froze all Japanese assets. The United States demanded that Japan withdraw its military and cease its aggression in China (#1). At the time, the Minister of War for Japan, Hideki Tojo, became the country's Prime Minister (#1). Roosevelt sensed the growing threat that Japan posed to its neighbors, so he sent General Douglas MacArthur to the Philippine islands on July 26th, 1941, to prepare for an assumed Japanese attack on Thailand, the Philippines, Malaysia, and other islands in that region. Japan invaded those places in order to have the vital resources it needed for military conquest.

On September 7th, 1941, Franklin's mother, Sara Delano, died (pg 272, #2). While mourning his mother's passing, Franklin had to deal with the cataclysmic international incident involving an American ship. The *U.S.S. Greer* was attacked by a German U-boat on, ironically enough, September 11th, 1941 (pg 277, #2). On September 19th, 1941, another ship, the *Pink Star*, was sunk by a German U-boat that was transporting a cargo of cheddar cheese (pg 282, #2). These acts of terror were followed by two more sinkings of the ships *Kearny* and *Robert James*, in which 100 more sailors were murdered.

On November 8ᵗʰ, 1941, the U.S. Senate passed the Neutrality Act, in order to arm ships in case of surprise attack. But by then, it was already too late to prevent the cowardly and devastating attack by the Japanese military that would catapult the United States into the Second World War. By November 3ʳᵈ, 1941, the Japanese knew that they were going to attack Pearl Harbor. As Japan became more aggressive in the world community, even Roosevelt feared that the Japanese military would strike at some point. However, he was expecting them to strike either the Philippines or Guam, and not Pearl Harbor. Even with these concerns taken into account, Roosevelt wanted Japan to strike first (#1).

On the morning of Sunday, December 7ᵗʰ, 1941, the United States experienced one of the worst moments in American military and terror history. That morning, the U.S. Navy had more than seventy-five ships—destroyers, cruisers, and submarines—stationed at the naval base at Pearl Harbor, and the Japanese were aware of this. The Japanese picked December 7ᵗʰ because it was a Sunday, and this atrocity would come as a complete surprise. Roosevelt himself proclaimed, to a joint session of Congress the next day, December 7ᵗʰ , 1941 to be "a day that will live in infamy" (#6).

Three hundred fifty Japanese pilots manned their planes in order to bomb Pearl Harbor, in a successful surprise attack (#6). All cadets in the Japanese military were taught to ask no questions of the orders issued to them, but just simply to obey them fully. The Japanese planned this attack for one year before executing this diabolical catastrophe. The Japanese wanted to have the biggest possible global audience to witness the attack.

At 5:15 a.m. on the morning of December 7ᵗʰ, 1941, the warplanes from the Japanese ships were ready to strike. They were located 220 miles off the coast of Hawaii. At 6:10 a.m., 183 planes were launched from Japanese ships. The Japanese plan of attack was to hit Pearl Harbor in three different waves, so that they could strike specific targets. At 7:02 a.m., the U.S. Navy picked up Japanese planes on radar, but failed to act. This was blamed on the military personnel's inexperience with radar equipment, and also on not recognizing the danger that these planes posed to the fleet at Pearl Harbor.

At 7:35 a.m., the mountains blocked the signal that would have detected Japanese air presence. At 7:52 a.m., the Japanese air fighters communicated to one another, "Tora, Tora, Tora," which meant success. The first wave of Japanese planes hit every American target imaginable. At 7:58 a.m., the second wave struck. The second wave of bombers hit military targets. The U.S. military tried to fight back but was caught by surprise in the confusion. The Japanese attacked the U.S.S. Arizona with torpedoes, sinking the mighty battleship in nine minutes; 1,177 men onboard the U.S.S. Arizona were killed. U.S. soldiers and Naval men who were being attacked that morning thought the attack was coming from

Hitler (#6). Even some American planes were shot down by American gunners on the ground, due to being confused about their targets.

At 9:05 a.m., a warning was dispatched by the American military base in Pearl Harbor for civilians not to go outside. The Japanese attack far surpassed their own expectations for successfully catching the Americans unprepared. The attack lasted for one hour and fifteen minutes. By 10:00 a.m., the Japanese attack planes were returning to their ships. By 1:30 p.m., the Japanese called off the third attack wave, due to imprecise coordinates for the third area of assault. Only twenty-nine Japanese planes out of 350 were shot down by American artillery.

By 12:10 p.m., despite the fact that the Japanese attack was over, American planes reported to patrol the area, in preparation for what seemed to be an inevitable land assault. American forces raided the Japanese embassy by 12:30 p.m. But the Japanese military campaign went far better than expected on the first two assaults, so the Japanese felt that a third strike would be unnecessary. Of the 461 servicemen who were still trapped inside the ship *Oklahoma*, only thirty-two made it out alive. Hawaii then went into an immediate state of martial law, and Japanese citizens were detained and sent to isolation camps. This became another controversial decision by the Roosevelt Administration. However, due to the shock and magnitude of the devastation at Pearl Harbor, Roosevelt felt that he had to take extreme measures to protect America.

In total, 2,403 U.S. servicemen were killed on the morning of the attack, and 1,110 more were wounded. Half of the fatalities were from the *U.S.S. Arizona*, on which 1,177 men were killed on that awful morning (#1). Out of the ninety-six U.S. ships stationed at Pearl Harbor on December 7th, 1941, nineteen were sunk, including four battleships: *Arizona, California, West Virginia,* and *Oklahoma.* The U.S. Navy also had ninety-two aircraft destroyed and thirty-one damaged. The U.S. Army had ninety-six destroyed planes and 128 that were damaged. American workers labored around the clock to rebuild the damaged material. Of the nineteen destroyed ships, fourteen of them would be rebuilt and went on to fight in the war (#6).

Consequently, on December 8th, 1941, President Roosevelt formally asked Congress for a Declaration of War versus Japan (#1). America was now thrust into a war that it had sought to avoid (pg 289, #2). This was considered the most serious military challenge since the Civil War (pg 292, #2). The Japanese military also attacked Malaysia, Wake Island, Guam, and Hong Kong on December 7th, 1941 (pg 295, #2). By December 11th, 1941, Germany had officially declared war on the United States, forcing President Roosevelt to also declare war on Hitler and Mussolini.

Yet there were only 524 American troops defending the Wake Islands, who sank two Japanese destroyers and killed 5,000 Japanese soldiers. In response, the Japanese sent reinforcements to the Islands, and they conquered the Wake Islands two weeks after Pearl Harbor. By the end of

December 1941, Japan controlled Guam, Hong Kong, and the Wake Islands. The fear of an attack by Japan on the coast of California produced an anti-Japanese political climate, despite them having nothing to do with the attack. This anti-Japanese atmosphere was sometimes quite unjust. Yet as First Lady Eleanor Roosevelt had noticed by December 15th, 1941, the world had indeed been transformed (pg 298, #2). The United States was no longer an isolated country and was a joint partner in the Allied war campaign.

The United States was united in participating in the war after the Pearl harbor attack. Churchill would often visit Roosevelt in the White House during this time for strategic discussions (pg 301, #2). Because of America's entry into the war, it increased its "lend-lease" program of sending supplies to Allied countries (pg 304, #2). The U.S. submarine fleet attacked Japanese shipping in order to hurt Japan's economic foundations. In fact, a Japanese submarine actually shelled the California coast, the first foreign nation to do so since the War of 1812 (#1).

On the Eastern Front, Russia clearly needed military and humanitarian aid, due to the vicious onslaught it was sustaining from Germany. For example, one million people died in Leningrad when the 1900-day siege ended; 3,000 people died each day from starvation as Germany advanced into Russia. However, the brutally cold winter seriously hurt the Germans and would eventually help Russia turn the tide to victory against the Nazis (pg 304, #2). Thus, the war coalitions were formed, pitting the Allied powers (America, England, and Russia) against the Axis powers (Germany, Italy, and Japan).

Churchill successfully lobbied the U.S. Congress for full support of the war venture (pg 309, #2). China also sided with the Allied powers. The Allied forces wanted construction of 60,000 warplanes, 4,500 tanks, and 20,000 anti-aircraft guns for use in this conflict (pg 313, #2). However, January 1942 was not a good start for the Allied campaign. Forty-three Allied ships had been sunk in a month's time, and more than 1,000 U.S. servicemen had died (pg 317, #2).

On February 15th, 1942, Japan beat Britain badly at the Battle of Singapore. Not only did Japan capture the territory, but it also captured 130,000 British soldiers there. In order to reassure the American people and the Allies that the war had become a morally just cause, President Roosevelt explained what needed to be done to beat the Axis powers on February 23rd, 1942. Roosevelt compared the war to the Revolutionary conflict and examined Washington's overwhelming upset of the British in the cause of what was right (pg 319, #2). Roosevelt further described a new kind of war, which would be waged on "every continent, every island, and every sea" (pg 320, #2).

However, with the fall of the Philippines a foregone conclusion, Roosevelt ordered General MacArthur to Australia in order to regroup. Yet this left thousands of U.S. troops vulnerable to be captured by the

Japanese (#1). It was then that MacArthur made his famous statement, "I shall return" (#5). Unfortunately, thousands of U.S. soldiers were captured by the Japanese at Bataan in April 1942. This was the largest army in American history to surrender to the enemy. Barbarically, the Japanese executed many U.S. military prisoners immediately. Next, the surviving soldiers were ordered by the Japanese on a 65-mile forced march that resulted in the murder of more than 5,000 U.S. soldiers. Many U.S. prisoners were beheaded or bayoneted. The Japanese also killed anyone at random who could not keep up with the others (#5). Some people blamed General MacArthur for the surrender at Bataan. Yet MacArthur would get his revenge by liberating Bataan two years later.

The U.S. war in the Pacific was well under way by April 18th, 1942. Roosevelt decided that the war had to be fought systematically in order to defeat the Axis powers. Though Japan was and continued to be a serious problem in the Pacific theatre, the Allies decided to take out North Africa first, because it was controlled by the Germans. This was done to distract the Nazis' vast military operational plan. President Roosevelt authorized a plan of incredible American mass production of tanks, planes, and weapons. Producing 800 planes a month, America's productivity far outstripped Hitler's, as German production was estimated at approximately 15% of the U.S. numbers (pg 363, #2).

On the home front, gender barriers were being broken, with women being given jobs in the war production department and jobs in essential civilian sectors. In fact, because of the fantastic accomplishments of women factory workers during the war period of 1940-45, 12,677 B-17 bombers, also known as "air fortresses," were created for Allied military use. Women workers had a lot to do with both the U.S. and the Allied success in the war, and this initiative provided jobs to grow the U.S. economy. The employment number for women grew to 80% (pg 381, #2). African-American women's employment opportunities grew during the war as well, and they were making better money during Roosevelt's tenure as president (pg 370, #2).

In addition to these numbers, four million men were added to the U.S. Army, raising the total number of men from 1.6 million to 5.4 million (pg 372, #2). By the end of 1942, unemployment had virtually been eradicated in the United States, and millions of Americans advanced above the poverty line (pg 399, #2).

In January 1943, Roosevelt met with Prime Minister Churchill at Casablanca in northern Africa for an Allied policy discussion. Roosevelt was the first president to fly overseas, and the first president since Abraham Lincoln to visit U.S. troops in an active theatre of war (pg 401, #2). It was in Morocco that both leaders concurred that the Axis powers must unconditionally surrender to end the war (#1). Roosevelt discussed with Churchill the Stalingrad battle versus Germany, which cost Russian forces more than one million fatalities. Roosevelt responded by sending heavy supplies to Russia to aid the country as much as possible. Roosevelt wanted to invade Italy before France, which was in contrast to Churchill's idea to invade France first, because Germany was more of a threat to England there. Roosevelt's task was to convince Churchill of his ideas, and he was successful. Ultimately, he would be proven right.

By July 10th, 1943, the Allied invasion of Italy had begun (pg 448, #2). Allied forces were led by General Dwight Eisenhower, who was appointed Supreme Commander of Allied Forces in 1943. Eisenhower's forces captured 275,000 enemy troops. The campaign took six months to complete. U.S. casualties in this campaign were 10,820 killed, 39,575 wounded, and 21,415 missing/captured (71,810 casualties) (pg 100, #11). General Eisenhower would not only go on to win the war, but would also later win two consecutive presidential terms in 1952 and 1956. It was Eisenhower who said, "The only unforgivable sin in war is not doing your duty" (pg 98, #11). Eisenhower personally hated war, but he loathed the Nazis more. Thanks to Eisenhower's military leadership and Allied military victories, by July 28th, 1943, Mussolini was ousted from power in Italy.

Nearly 250,000 German and Italian soldiers were taken prisoner by Allied forces in North Africa (pg 437, #2). Meanwhile, in the Pacific war theatre, the important Battle of Midway had begun on June 4th, 1942 (#1). This followed the U.S. military victory at the Battle of the Coral Sea against Japan, which was a morale boost for the Allies. Even though the Japanese inflicted substantial damage to some U.S. ships, the U.S. military inflicted even more damage and consequently won the battle. The U.S. military lost 547 men at Midway, one destroyer, and 150 aircraft, and the Japanese lost four battleships, 360 planes, and 3,500 men (#1). The Battle of Midway would alter the course of the war in favor of the United States.

Next, the U.S. wanted to retake Guadalcanal, which the Japanese had captured. The first month of fighting was extremely fierce and brutal, but the United States was winning. The United States and the Allies successfully reclaimed the island at the cost of 1,600 U.S. soldiers. The Japanese losses were estimated to be as high as 15,000 soldiers (#1). The Allied successes were credited in part to the outproduction of planes by three to one. Hitler, unlike Roosevelt, did not consider employing women to produce war-making material. However, Hitler did utilize captured Russian prisoners as slave labor.

Meanwhile, Hitler gave the order to continuously kill Jewish people in concentration camps, because Hitler was consumed with evil. The Germans also killed 2.8 million of the 3.9 million prisoners they had captured. Yet the Soviet military fought Hitler ruthlessly on the Eastern Front and killed many German troops there. Soviet forces beat Hitler out of the Crimea and at Sevastopol, and were on the offensive (#1). By April 1944, Operation Persecution was launched by the Allies in order to recapture New Guinea in the Pacific theatre, which they successfully did (#1). General MacArthur and his forces had already invaded New Guinea in February 1944, and they advanced slowly towards the Philippines.

At this time, Roosevelt and Churchill began planning for an Allied cross-channel attack to take place against the Nazis in France, scheduled for around May 10th, 1944 (pg 460, #2). General Dwight Eisenhower was selected to lead this operation, with General Bernard Law Montgomery of England heavily involved in strategic planning. This plan was called Operation Overlord (#1). Germany could not rely on Japan, because Japan did not reinforce Germany militarily in any way, besides fighting America (#1). The Allies tried to trick the Nazis into thinking that the invasion was coming from Pas-de-Calais, which was the most direct route from France to Germany, and not from Normandy (#1). "D-Day" was rescheduled to begin on June 5th, 1944, but again had to be delayed a day due to inclement weather.

The American First Army was to land at Normandy Beach, the Fourth Infantry was to land at Utah Beach, the First and 29th Infantry invaded Omaha Beach, the British 50th Division invaded Gold Beach, and the Canadian Third Division attacked Juno Beach on the morning of June 6th, 1944; 200,000 soldiers invaded the 60-mile coast on the first day (#1). The British had very light casualties and resistance on Gold Beach, and were five miles inland by nightfall. The same story occurred at Juno Beach. However, at Normandy and Omaha beaches, the fighting was fierce: 2,500 soldiers were killed, out of 175,000 who landed on the first day of the invasion, and there were 6,600 casualties on the first day (pg 510, #2).

Despite the number of casualties, the Allied invasion inflicted some serious damage on the enemy. By three weeks after the invasion, one million Allied soldiers were ashore, 171,532 vehicles had landed, along with 556,000 tons of supplies. The Allied intent from this point was to invade France and continue all the way to Berlin to force a Nazi surrender. U.S. airpower was fighting virtually unopposed. Teddy Roosevelt, Jr. was the only brigadier general to land with his troops on Normandy Beach on June 6th, 1944, and later died of a heart attack on July 13th, 1944. He was 57 years old and is buried next to his brother, Quentin, who died during World War One (#1).

To honor the memory of those who sacrificed their lives in the field of battle on D-Day, Roosevelt signed the G.I. Bill of Rights on June 22nd, 1944. This provided status, education, and training to G.I.s returning from the war. The United States continued to supply the Soviet Union with war material, in return for Soviet help beating Japan after Germany surrendered (pg 477, #2). By July 1944, Normandy had surrendered to the Allies, and Russia had invaded the German-controlled territory on the Eastern Front.

With Hitler now clearly losing the war, he ordered the round-up and extermination of as many Jews from western and central Europe as possible (pg 515, #2). However, 600,000 American Jews served bravely in the armed forces during World War Two, including future National Security Advisor and Secretary of State Henry Kissinger (#1). To deter the persecution of Jews in Europe, Allied forces started bombing rail lines to disrupt the transport of the Jewish prisoners, but Roosevelt's staff reacted way too slowly on this issue, because they were more focused on beating the German military first.

By August 26th, 1944, the Allies recaptured Paris. General Montgomery of England then liberated Brussels (#1). There were a reported 130,000 German fatalities during the last week of June 1944, and 6,600 German soldiers who were taken prisoner. On Tuesday, July 11th, 1944, Franklin Roosevelt announced his intentions for running for an unprecedented fourth term as president in the 1944 presidential election. Roosevelt's platforms for his presidential campaign in 1944 were as follows:

1. Win the war
2. Form a worldwide international organization
3. Build the economy for returning veterans and for all Americans (pg 529, #2)

Because Vice President Henry Wallace was not that popular among fellow Democrats, Roosevelt chose to leave the position of vice president up to a vote among Democrats at the convention. They chose Senator Harry Truman from Missouri, who often agreed with Roosevelt's presidential policies. He also voted accordingly. Roosevelt chose Truman over such candidates as James Byrnes, Sidney Hillman, and Wallace, because Truman was the least controversial of the bunch, and he would be the most logical choice for replacement should Roosevelt not finish a fourth term. Truman won easily on the second ballot (pg 320, #7).

By July 1944, even the Germans were trying to kill Hitler (pg 531, #2). Hitler forced his commander, General Erwin Rommel, to commit suicide in order to avoid trial for treason for an attempt against Hitler's life (#1). In the Pacific war theatre, the United States battled Japan for control of Saipan, Guam, and Tinian, at the cost of 1,400 U.S. Marine lives. The Allies still won control of these islands, which were later used to bomb mainland Japan.

General MacArthur lobbied Roosevelt to let him liberate the Philippines himself. MacArthur desperately wanted to return to Bataan, to personally liberate the troops he left behind. Roosevelt granted MacArthur's request of 600 ships from the U.S. Navy for his return to Bataan. As MacArthur accurately predicted, he did return (#5). Sadly, only a third of MacArthur's men left behind actually survived the Bataan Death March. Roosevelt later promoted MacArthur to a five-star general.

Meanwhile, Roosevelt beat Republican presidential candidate Governor Thomas Dewey of New York on election night, November 7th,

1944, winning an unprecedented fourth term as president. However, Roosevelt won this election by the slimmest margin of his four victories: 25,602,505 votes (53%) to Dewey's 22,006,278 votes (46%). Despite this, the electoral vote heavily favored Roosevelt, who secured 432 electoral votes to Dewey's 99 (pg 493, #4). Roosevelt could boast that his accomplishments included the war victories, women's employment in war-related jobs increasing by 46%, and a 4-fold increase in female membership in unions (pg 555, #2).

Back in Europe, the single biggest battle on the Western Front, and the largest engagement ever fought in by the U.S. Army, was the Battle of the Bulge, on December 16th, 1944. The Battle of the Bulge was fought in the western German mountains where Belgium, Luxembourg, and Germany meet. Hitler ordered the attack, composed of mostly German teenagers. They attacked an area in the Ardennes where four U.S. divisions were stretched out over 80 miles. After this, Hitler had planned to go on to Antwerp, in order to control the port.

The Germans had 200,000 of these young troops and placed them in position to attack the Allies. Hitler thought that Eisenhower would have to ask the president for approval for a defensive plan, which would have come too late. However, Hitler was dead wrong. The attack began on December 16th, 1944, and the Germans made some advances. Yet some German divisions were decimated by the Allied forces from their fortified positions.

U.S. troops fought off advancing German troops at Elsenborn Ridge, though they were outnumbered five to one. Though the Allies were successful at Elsenborn Ridge, 9,000 Americans surrendered to the Germans at Schnee Eifel (pg 564, #2). German troops massacred captured U.S. troops on the second day of the Battle of the Bulge in the Belgian town of Malmedy, killing approximately ninety-six to 130 troops (#1). General Eisenhower sent the Seventh and 12th divisions to St. Vith to stop the Germans, and it worked. By January 1945, General George Patton and his Third Army reclaimed all of the territory that Germany had taken on the offensive, and the German tactical advance at the Bulge was literally erased (pg 565, #2). Patton's Fourth Armored Division broke through the city of Bastogne after some intense combat.

The Allies won this important battle, but the losses in human life were astronomical and horrible for both sides. Of the 600,000 U.S. troops involved in this battle, nearly 20,000 were killed, 40,000 wounded, and 20,000 captured. Yet 30,000 German soldiers were killed, 40,000 wounded, and 30,000 taken prisoner. African-American soldiers for the United States fought bravely at the Battle of the Bulge and were complimented for their heroism (pg 568, #2). After four weeks of fighting at the Bulge, the tidal wave of momentum was carrying the Allies to victory.

By early 1945, the Allies had freed Paris, Brussels, Warsaw, Belgrade, Budapest, and Athens (pg 578, #2). The Battle of the Falaise Pocket was the battle that ended the German occupation of France, at the price of 10,000 Nazi soldier fatalities and 50,000 captured. General Eisenhower was credited with the victory, but he complimented both Generals Montgomery and Patton for their tactical operating decisions and accomplishments.

On the Eastern Front, Russia invaded Poland on January 12th, 1945, and reclaimed that country from Nazi dominance (pg 817, #8). The Soviet Army would go on to capture Budapest and Vienna by April 13th, 1945 (#1). Stalin wanted territory reclaimed for Russia stemming from the losses during the Russo-Japanese War of 1905 (#1). Roosevelt was also contemplating the wishes of Russia to occupy Poland, because Stalin argued that both Napoleon and Hitler had used Poland to brutally attack Russia (pg 582, #2). The meetings between Roosevelt, Churchill, and Stalin at the Yalta Conference of 1945 cemented Roosevelt's dream of establishing a United Nations between countries (pg 583, #2).

Though Roosevelt was pleased with the military progress in Europe, he estimated the casualty list from a possible invasion of Japan to go as high as one million people (pg 580, #2). For example, on February 19th, 1945, the important Battle of Iwo Jima was fought, which claimed the lives of 70,000 Marines, with 19,000 men injured, but this battle was a very important victory in the quest to control the Pacific (pg 589, #2). The next battle was for Okinawa, which became the biggest battle in the Pacific, involving 548,000 U.S. troops. This was the greatest amphibious assault of all time (pg 827, #10).

By the spring of 1945, Roosevelt had secured his beloved country from the grips of totalitarian dictatorships abroad, and revitalized the nation's economy with his New Deal economic programs. As he was leading his country at the beginning of his fourth term as president, Roosevelt was still looking forward to a post-war world in which his nation would prosper and thrive and be free from the terror grip of the Axis powers. Then suddenly, on April 12th, 1945, Roosevelt collapsed in Warm Springs, Georgia from severe head pain. He was medically diagnosed with having a cerebral hemorrhage. President Franklin Roosevelt died on that day. Both Churchill and Stalin were devastated by the news (pg 606, #2). Vice President Harry Truman became the 33rd President of the United States that day (pg 345, #7).

Upon Truman taking the oath of office, he insisted that the war would not end until the Axis powers had "unconditionally surrendered." Under the late President Roosevelt's leadership, the United States built 300,000 war planes, two million trucks, 107,351 tanks, 87,620 warships, 5,475 cargo ships, 20 million rifles, and 44 million rounds of ammunition. Also, six million people joined unions during this time (pgs 607-608, #2).

Seventeen million new jobs were created between the years 1940 and 1945.

By April 1945, Hitler and his forces were in undeniable peril: 500,000 Russian soldiers, led by General Georgi Zhukov, invaded Berlin. Allied bombing had killed 600,000 German civilians in Dresden, Hamburg, and Cologne. Hitler and Eva Braun secretly got married and committed suicide in an underground bunker on April 30th, 1945. Nazi leader Joseph Goebbels killed his six children, his wife, and then himself also. Consequently, on May 7th, 1945, Germany surrendered to the Allies. The war in Europe was over, and the continent was free of Hitler!

Surprisingly, Winston Churchill was swept out of office on May 8th, 1945, when his party was beaten by the Labour Party, which then took over Parliament (pg 621, #2). Attacks by B-29 bombers in Japan on March 9th-10th, 1945, killed 100,000 people. Yet the Japanese Army continued its attacks on Filipino citizens, raping and killing them in 1945. This is similar to the rape of Nanking that the Japanese Army committed earlier in the war against China. Thankfully, the Philippines were recaptured in March 1945 by the Allies. The Japanese spirit of not giving up in battle made President Truman decide that he needed to end the war sooner rather than later, and some experts estimated that the conflict could stretch into 1947, with an additional 200,000 casualties.

The Japanese military forces were known as tremendous fighters, as evident after the Okinawa battle, which killed 12,000 U.S. soldiers and left 36,000 soldiers injured. America could save an estimated 500,000 to one million soldiers by avoiding a land invasion of Japan, but this would require the use of the atomic bomb. Once Truman had definite confirmation of the atomic bomb's viability, he demanded that Japan surrender, but they refused (#1). The first nuclear explosion in history was successfully orchestrated at Alamogordo Air Base in New Mexico on Monday, July 16th, 1945 (pg 416, #7).

Hiroshima was picked for the atomic bombing by the United States because it was the Japanese southern army headquarters, and the depot for Japan's homeland army (pg 436, #7). Truman's decision to drop two atomic bombs on Japan was, in his mind, the only way to get Japan to surrender and to save as many American and Japanese lives as possible. Some leaders, including General Eisenhower, believed that Japan had already lost the war, but President Truman saw that the casualties were still climbing in the Pacific war theatre. Roosevelt had definitely approved the Manhattan Project idea, and everyone felt that he would have used it under the same circumstances. Truman was also sympathetic to the U.S. ground forces who would have been lost in a mainland invasion of Japan, as he was once an artillery officer in World War One (pg 440, #7). Yet Truman was acutely aware that nuclear weapons could lead to global armageddon (pg 518, #10).

Consequently, President Truman ordered the atomic bombing of the city of Hiroshima, Japan, which commenced on August 6th, 1945. Approximately 250,000 people were either instantly killed in this explosion or died of radiation exposure within a year. Despite this, Japan still refused to surrender, so a second atomic bomb was dropped on the city of Nagasaki on August 9th, 1945. These devastating attacks brought Japan to consequently surrender to the Allies on August 15th, 1945. This brought World War Two officially to an end.

The overall statistics of one of the world's most horrible conflicts are staggering. The total fatalities of people who were murdered during this war numbered approximately 57 million. Germany lost 3.2 million soldiers and 3.6 million civilians in this war, while Japan lost two million civilians and one million soldiers. Six million European Jews were murdered at the evil hands of the Nazis.

Of the Allies, the Soviet Union suffered the heaviest losses in this conflict, with 7.5 million fatalities (pg 518, #10). Though President Truman initially found Soviet leader Stalin favorable to the Allies, Stalin's humanitarian record in Russia was atrocious. Stalin was responsible for the deaths of five million Russian peasants who opposed him, and 10 million more Russians were forced to work in labor camps (pg 419, #7). Stalin was one of the worst mass murderers of all time, comparable to Hitler or Ivan the Terrible.

Even though Italy had been on the Axis side during World War Two, President Truman thought that it should be included in the United Nations. In fact, it was Truman who led America in the creation of the United Nations. Truman was instrumental in creating the nation of Israel in 1948. This stemmed from the belief that Jewish people deserve a homeland. Truman also created the Marshall Plan for post-war Europe, named after Secretary of State George Marshall, which successfully rebuilt Europe from the ashes of the war. During 1948-52, the United States spent some $13 billion on reconstruction projects, and pulled the European economy back from the abyss and into prosperity and normalcy (pg 519, #10).

Truman saw the need for interaction in third-world countries in order to promote improvement and growth in Africa, Asia, and Latin America. The Nuremberg Trials of 1945 and 1946 saw the prosecution of twenty-two Nazi commanders, who were on trial for their horrendous crimes against Europeans, especially the Jewish people. Nineteen of the twenty-two men tried were found guilty, and twelve of them were executed, including Hermann Göring.

Future Presidents Dwight Eisenhower, John Kennedy, Lyndon Johnson, Richard Nixon, Gerald Ford, Jimmy Carter, and George H.W. Bush (#11) had all served bravely in this brutal conflict, and its ramifications affected how all of these men viewed foreign policy in the United States. The future conflicts during the Cold War were shaped on the Truman Doctrine to stop the Communist advancement on free countries, and he sent aid appropriations to Greece and Turkey to stop the Communist influences in those governments in 1947, which proved successful. Truman also created the Central Intelligence Agency (C.I.A.) in 1947. The United States and the Allies successfully stopped the Axis power aggression of Germany, Italy, and Japan and rebuilt those countries into democratic and peaceful institutions. Both Presidents Franklin Roosevelt and Harry Truman deserve much credit for making the extremely difficult decisions that sent our soldiers into harm's way, and for formulating a successful outcome for democracy everywhere, though coming at the price of just under 300,000 American soldiers' lives (pg 348, #2). General MacArthur was sent by President Truman to help the Japanese people rebuild their lives under a successful democracy, and MacArthur would accomplish just that. All would be well until another conflict erupted in 1950 in the Asian country of Korea, where Communism was spreading at an uncontrollable rate.

Bibliography

1. *The Second World War,* ed. by John Keegan (Penguin Books, 1989)
2. *No Ordinary Time*, by Doris Kearns Goodwin (Simon & Schuster, 1994), disc 4.
3. *Roosevelt: The Lion and the Fox* (1882-1940), by James MacGregor Burns (Harvest Books, 1984), disc 8.
4. *The Complete Book of U.S. Presidents,* by William A. Degregario (previously listed).
5. PBS home video: "MacArthur: The American Experience," disc 1.
6. Learning Channel video: "Pearl Harbor: Seven Views of Defiance."
7. *Truman*, by David McCullough (Simon & Schuster, 1992), disc 4.
8. *Okinawa: The Last Battle of World War Two*, by Robert Leckie (Viking, 1995), disc 8.
9. *World War Two*, by H.P. Willmott, Charles Messenger , and Robin Cross (Dorling Kindersley, 200$), disc 1.
10. *The Wars of America*, by Robert Leckie (previously listed).
11. *George Bush: Life of a Lone Star Yankee*, by Herbert S. Parmet (Scribner, 1997), disc 5.
12. *Eisenhower: Soldier and President*, by Stephen E. Ambrose (Simon & Schuster, 1990), disc 5.
13. *Churchill and Roosevelt at War: The War They Fought and the Peace They Hoped to Make*, by Keith Sainsbury (New York University Press, 1994).
14. *Crusade in Europe*, by Dwight D. Eisenhower (Doubleday, 1948).

INTERVIEW WITH DOCTOR ROBERT KEIGHTON

Thomas Athridge: If World War Two was not on the horizon for the United States, would FDR have run for a third term as president?
Robert Keighton: Yes, it was an excuse to do so. I think that he would have returned because he felt that he had to run.
TA: Had FDR not been caught cheating on his wife, Eleanor, in September 1918, would Eleanor have had the impact as First Lady that she later had?
RK: Eleanor found letters between FDR and Lucy Mercer, and she was deeply hurt. Though divorce was publicly frowned on at the time, Eleanor did consider it. Though they eventually reconciled, Eleanor did have a mind of her own, and she became more self-confident and developed strong views. She had enormous influence on FDR, who was more cautious in stating his views comparatively. Marian Anderson, who was black, wanted to give a concert at Constitution Hall. Because of her color, she was denied her request. Eleanor protested this decision and switched the venue to the Lincoln Memorial, where she performed. She revealed the conditions of normal people that FDR did not know about.
TA: Did Germany's invasion of Europe help FDR with the U.S. economy by selling weapons to our European allies?
RK: We did not get out of the depression in the 1930s. Workers did gain economic value with weapons construction, and it did help the United States out a lot.
TA: How off-guard was FDR caught when the Japanese attacked Pearl harbor?
RK: Yes, Pearl Harbor was a total shock when it happened. The feeling of the time was that it was foolish of the Japanese to do this, because they could have grown stronger elsewhere in the world. At first, the United States concentrated on Germany. In the long haul, this decision was a disaster for Japan. It galvanized the American people to fight. We concentrated more on Japan after 1944.
TA: Was Pearl Harbor similar enough to the 9/11 attacks on America, though the Japanese attacked a military installation?
RK: Yes, because it had the same effect. The U.S. people were completely behind the effort to punish those responsible.
TA: Was FDR justified by putting Japanese Americans in isolated camps?
RK: FDR was mistaken in putting the Japanese in these camps: 110,000 of them were wrongly placed there. Ironically, German and Italian people were not placed in these camps. This is a bad blemish on FDR's record.

TA: Did FDR's revolutionary campaign to employ women during wartime give the United States the advantage in the war?

RK: Yes, it did contribute to the victory, because this increased the ability to manufacture weapons. It did help employment for people, which in turn helped the economy.

TA: How revolutionary was the G.I. Bill of Rights for returning veterans?

RK: Yes, this was important because it provided a college education and gave them a leg up on high-ranking positions.

TA: How important was the Allied invasion of D-Day, June 6, 1944, in the Allied victory versus Germany?

RK: This invasion was very important, and a very dramatic moment in the direction of the war. The Russians were waiting for the Allied invasion and wondering why the invasion was not sooner.

TA: In your opinion, was the cause of World War Two the failure of the meetings of Allied leaders who signed the Treaty of Versailles in 1919, at the end of World War One?

RK: Oh yes, I think so. Germany was being asked to pay too much money to the winning countries. The economic conditions were too horrendous for Germany to comply with. After World War Two, the Marshall Plan was a much more stable plan to comply with.

TA: Why did FDR's dream of a United Nations work, while President Wilson's idea for a League of Nations failed?

RK: The United States did not join the League of Nations, and we were more active in the world after World War Two. Before World War Two, the United States felt isolated and secure, but after Pearl Harbor, we did not.

TA: Did FDR's successor, Harry Truman, make the correct decision in dropping two atomic bombs on Japan, as opposed to sending U.S. troops into Japan for control of that country?

RK: Truman did in fact make the right decision, although it was a tough one. He made this decision based on the Battle of Okinawa. But once Truman made up his mind, his decision was firm. Truman also got more credit for integrating the Armed Forces.

Chapter 11
The Korean War (1950-53) and the Decisions of Both Presidents Harry Truman and Dwight Eisenhower

The American military involvement in the Korean War began on June 25th, 1950, after North Korea had invaded territory south of the 38th Parallel. Korea was under the control of the Japanese, who had conquered it during World War Two, but had regained its territory at the war's conclusion in 1945. At the time of the war's end, Korea was occupied in the north by the Soviet Union and in the south by the United States.

President Truman did not give South Korea any military weapons, for fear that the South Korean government would unnecessarily attack the north. However, the reality was that North Korea was being readily supplied by the Soviet Union with all of the modern weapons of the time. Also, from 1945 to 1950, the North Korean Army grew to 136,000 troops. North Korea waited for the chance to attack and take over the south, which occurred on June 25th, 1950.

On that day, seven North Korean divisions attacked the south with Russian artillery to support them. The South Korean Army was outnumbered and outmatched, with inferior firepower. President Truman was not even told about the Communist invasion on the 38th Parallel until nine hours after the fact. Because the newspapers knew about the invasion in Korea before Truman did, the President was extremely angry. He knew that North Korea's actions of hostility against its neighbor would certainly motivate American involvement in this war. But in order to understand why President Truman made the decision to fight an aggressor overseas while putting American soldiers' lives at risk, it is important to understand the background of our nation's 33rd President of the United States, and why he felt that it was so important to defeat a Communist enemy in a world still reeling from the effects of the Second World War in a far-away land in Asia known as Korea.

Harry Truman was born on May 8th, 1884, in Lamar, Missouri (pg 37, #2). Truman was born in a racist and segregated society, where lynchings of black people were common. Truman's father, John Anderson Truman, was a farmer and livestock trader who suffered a severe financial setback by losing all of his money after investing in a bad crop of wheat in 1901. This incident occurred during Harry's senior year at Independence High School. At this time, Harry applied to West Point but was rejected (pg 67, #2). As a result, Harry did not attend any school and became a bank clerk.

Truman would go on to become a successful banker, deacon, and farmer before he joined the military in 1917 to fight in World War One. Truman had many legitimate excuses not to go to war, including having bad vision, being past the draft age, and being the sole financial supporter to his mother and sister. Despite these facts, Truman chose to serve in the Armed Forces anyway. Truman felt strongly that "we owed France something for Lafayette," and he believed strongly in the Allied cause.

The slaughter of people in the war before America's entry in it during April 1917 can only be described as barbaric. In 1916, there were an estimated two million casualties on the Western Front. In the four-month Battle of the Somme (July-November 1916), Germany lost more troops than the United States did during the entire Civil War. When the German/Austria-Hungary aggression finally provoked the United States to enter the war, the U.S. Army grew from 14,000 to two million men during its peak. Truman's own conduct in the war campaign won him praise from his troops and superiors. He was described as having great personal conduct and strength as a leader of men. Praise for Truman included recognition for saving the lives of his men during battle, and for making wise decisions to avoid capture or slaughter.

Truman became widely admired and commended for his military service. Truman's lifelong heroes included Andrew Jackson and Robert E. Lee (pg 141, #2). Upon Truman's return from the war in Europe, he married his wife, Bess, on the day the Treaty of Versailles was signed, on June 28th, 1919 (pg 510, #3). He was also convinced by his friend Mike Pendergast to enter the political world by running for a vacant judicial position in 1922, and Truman won the position. Truman remained a judge until January 1935. Truman's duties included handling an operating budget of over $7 million, which he did successfully.

A vacant U.S. Senate seat opened in the state of Missouri in 1934. Truman decided to campaign for and pursue the Senate job, and he was victorious in the election. Once Truman was on Capitol Hill, he was vocal in his beliefs that the United States was mistaken at the end of World War One by not signing the Treaty of Versailles, and by not joining the League of Nations. This would influence Truman's decisions later when he became president.

During this time, President Roosevelt was sending Britain supplies and money to fight Germany. Roosevelt proclaimed that people around the world should be guaranteed four freedoms:

1.   Freedom of speech

2. Freedom of religion
3. Freedom from want
4. Freedom from fear

These rights are still held true in today's democracy. When Truman advanced to the U.S. Senate, he often found himself in agreement with President Roosevelt and his policies, and voted accordingly. During Roosevelt's unprecedented fourth straight campaign for president in 1944, Senator Truman's name was mentioned for the vice president slot on the ticket, but names like Henry Wallace were the more likely picks. However, in a surprise move to some, Roosevelt did pick Harry Truman. This decision was made by President Roosevelt because, in his mind, Truman filled two roles that satisfied the president:

1. Truman was the least controversial of the candidates Roosevelt had selected.
2. Truman would be the most logical choice to replace Roosevelt should he not be able to finish a fourth term.

Truman claimed victory on the second ballot at the Democratic Convention of 1944. Roosevelt and Truman would go on to win the presidency and vice presidency on November 7th, 1944 (pg 492, #3).

Though the struggle against the Axis powers had started to turn in favor of the Allied side, Roosevelt was visibly growing more sick as 1945 began. On April 12th, 1945, President Roosevelt died, and Vice President Harry Truman became the 33rd President of the United States. When he took office, Truman insisted that the war would not end until the Axis powers had "unconditionally surrendered." On May 7th, 1945, Truman got exactly that from Germany. Hitler and Eva Braun had already committed suicide together on April 30th, 1945.

In the Pacific war theatre, American B-29 attacks in Japan on March 9th-10th, 1945 killed 100,000 people. However, the Japanese were known for not giving up in war, and also for being tremendous fighters. Truman knew that the only way to force the Japanese military powers to the negotiating table was the use of the atomic bomb. Truman decided that in order to save American soldiers' lives—and the lives of the Japanese, for that matter—and end the war as quickly as possible, which some experts estimated could last as long as 1947, with an additional 200,000 U.S. casualties inflicted in a land invasion of Japan, the use of the bomb was absolutely necessary.

The first explosion of a nuclear bomb in history was successfully orchestrated at Alamogordo Air Base in New Mexico on Monday, July 16th, 1945 (pg 416, #2). Hiroshima was picked as a target to strike with a nuclear weapon because it was Japan's southern army headquarters and the depot for Japan's homeland army. President Truman felt that this was the only way to get Japan to surrender and successfully end the war. When a second bomb was dropped on Nagasaki, Japan on August 9th, 1945, Japan had no choice but to formally surrender. They would officially do so on the battleship *Missouri* on August 15th, 1945, effectively ending World War Two.

When the war was officially over and the Allies had won, President Truman enjoyed the victory, but his joy would not last long. Though America and the Allies had defeated Fascism in 1945, Truman would say, "Sherman is wrong, peace is hell." President Truman was thrust into power to lead a country that was celebrating victory over the most powerful combination of enemies that this nation has ever faced, and he was overwhelmed by the adulation. Twelve million victorious G.I.s came home after the war, and waited for opportunities to go to college and succeed in the federal government. However, at this time, labor and business were at each other's throats.

While Truman was not in control of the pricing, he was blamed for not doing enough to solve the economic crisis. To help the economy, Truman proposed increases in the minimum wage, as well as housing aid. Truman also proposed a bill for the first prepaid medical insurance in the history of the country. However, the Republicans, and some Democrats, rejected all of Truman's proposals. Truman compared being president to riding a tiger: either hang on or get eaten.

To make matters worse, 1946 brought 5,000 new strikes to America. President Truman desperately needed to have labor support as a Democrat, but many were angry at him and he at them. Truman loathed strikes, and that is exactly what the railroad employees did in May 1946, which crippled an important transportation industry and the mobility of the American public. In response, President Truman threatened to draft the strikers into the Armed Services. In fact, President Truman went so far as to appear before a joint session of Congress while appealing against the strike, to ask permission to implement this draft of striking men.

In a surprising development, this tactic ended the railroad strike literally in the middle of Truman's speech in front of Congress, which inspired the members to applaud from the chamber. Although this was a triumph for President Truman, he paid a high price for this victory. It cost him labor support, which would prove disastrous for the Democrats during the mid-term elections of 1946. During this time, Truman's popularity as president hit rock bottom. The phrase "To err is Truman" was popular at the time. Inflation was out of control, also.

The Republicans won both houses of Congress in 1946, and Truman was frustrated in implementing his policies and vision against a hostile House and Senate. Truman's decisions in Europe, however, made a 50-year impression on world affairs. Stalin and Truman admired each other at one time, but soon found themselves at odds over the Soviet occupation of Eastern Europe. While Stalin was in power in the Soviet Union, they had a mighty military. Truman said that he "grew tired of babysitting the Soviets," and saw the threat of war with them as being very real. Stalin, too, thought that war with the United States was inevitable.

Greece and Turkey became a turning point in the escalation of tension between the Soviet Union and the United States in 1947. The Communist Party of the Soviet Union threatened to take over both nations' governments at that time. President Truman chose to intervene against Soviet influence in both Greece and Turkey. Truman asked the Republican Congress for $400 million to save Greece and Turkey from becoming Communist nations. Truman phrased the request such that if a member of Congress were against the request, then it would mean that the person would be voting for slavery over freedom.

Truman also had help from Dean Acheson, who influenced members of Congress to support the measure. The result was that Congress approved Truman's request for the $400 million aid package to Greece and Turkey to fight the spread of Communism. This resolution would be forever known as the "Truman Doctrine." This act began the Cold War.

After World War Two, Western Europe was literally starving. In an effort to send aid, the United States sent $13 billion to feed Europe. Truman succeeded in his plan to save Europe, and he called this the Marshall Plan, named for his Secretary of State, George Marshall. Truman created several very important agencies while he was president: the Department of Defense, the Central Intelligence Agency (C.I.A.), and the National Security Council. Furthermore, N.A.T.O. (North American Treaty Organization) was established to unite America with free European nations to fight Communist aggression. The Cold War would last for nearly half a century.

Truman wanted to redeem himself and the Democratic Party in the elections of 1948. Racism was everywhere in those days, but Truman expressed outrage about it. In fact, President Truman was the first president to address the National Association for the Advancement of Colored People (N.A.A.C.P.). Truman wanted equal rights for every person in the United States, and acted on this by desegregating the armed forces.

Palestine was a foreign policy matter that President Truman dealt with regarding a homeland for the Jewish people. After the Holocaust, which killed more than six million Jews in Europe, the Jewish people asked for a homeland in Palestine. Truman sympathized with the horrors the Jewish people went through during World War Two, and on May 14th, 1948, Truman officially recognized the nation of Israel. While Truman believed that the Jewish people should have a homeland there, this act brought about the conflict between the Jewish nation and their Palestinian/Arab neighbors that has lasted for over 65 years.

Governor Thomas Dewey from New York was Truman's Republican opponent in the 1948 presidential election. Dewey was thoroughly beating Truman in the polls, and everyone assumed that Dewey would handily beat Truman for the presidency. But Truman seemed unfazed by the predictions, and he campaigned vigorously for re-election, criss-crossing the country three times. On election night, November 2nd, 1948, Truman successfully shocked the entire world by upsetting Governor Dewey, proving everyone wrong by winning the presidential election.

After the beginning of his second term, Truman pushed his "Fair Deal" legislation on Congress. In it, Truman requested an increase in the minimum wage, health care coverage for everybody, and funding for education and civil rights. However, in August 1949, everything changed when the Soviets exploded their first nuclear bomb. Only weeks later, China fell to Mao Tse-tung's Communists. Truman was urged by advisers to spend more federal money on defense, but the president found it more important to have a balanced budget.

Then the world and its events changed once again on June 25th, 1950, when the invasion that would lead to the U.S. involvement in the Korean War began as North Korean forces crossed the barrier of the 38th Parallel demarcating Soviet and American territory after World War Two. When Truman was informed about the invasion, he thought Stalin and the Soviets were involved in the Korean invasion. As a result of the Communist invasion of South Korea, President Truman ordered immediate retaliation. Truman would not back down to the Communists.

After he was almost defeated by Thomas Dewey for the presidency in 1948, and noting the growing tide of Senator Joe McCarthy's anti-Communist rhetoric on Capitol Hill, President Truman saw North Korea's advance through the 38th Parallel and into the south as an unacceptable act of aggression. This act led to further acts of hostile, Communist takeovers of other governments. Truman sought to prevent this by deploying U.S. combat troops to Korea. Truman compared the North Korean acts of aggression to that of the Kaiser of Germany during World War One, and to those of Hitler and Japan during World War Two, when America sat back and watched these countries grow stronger and more hostile. These situations developed until conflict was unavoidable between America and these nations.

While General Douglas MacArthur was in Japan after World War Two, helping with that country's reconstruction, he initially felt that there was little he could do to help the Korean situation unless he was sent by the president directly to Korea. Consequently, this is what he asked of the president. Truman granted MacArthur's request, and MacArthur went to Korea. United Nations resolutions also condemned the attack by North Korea against the south, and 16 nations sent troops to fight together. As a result, Truman increased the federal defense budget.

By the end of 1950, there were over 4,000 U.S. soldier fatalities and over 14,000 wounded G.I.s in Korea. MacArthur, now in military command, wanted permission from President Truman to invade Inchon, Korea on September 15th, 1950. Truman agreed, and MacArthur's military plan worked. Inchon fell in a day. Thirteen days later, Seoul was recaptured by the South Korean/American forces. By this time,

171

MacArthur was 70 years old and was serving in his third wartime scenario. Truman and MacArthur were tense with each other at times because Truman occasionally felt that General MacArthur would disobey orders.

Nevertheless, MacArthur's presence in military command seemed to make all of the difference. As a result of these American victories, MacArthur wanted to pursue the Communists past the 38th Parallel into North Korean territory, in an attempt to liberate the entire country of Korea. Truman gave the go-ahead to MacArthur to commence with this plan.

At first, Truman had the public's support for the war. He was seen as tough, smart, and decisive (#1). An invasion of North Korea seemed logical at the time. The U.S. military invaded the north, under General MacArthur's direction, and captured the North Korean capital city of Pyongyang on October 9th, 1950. Many people believed that the war would soon be over. MacArthur made it clear that he wanted to press on to China, which turned out to be a bad tactical decision.

This act officially provoked China's entry into the war, with 180,000 Chinese soldiers sent into Korea to confront U.S. soldiers at Koto-Ri City, Hagaru-Ri, and Yudan-Ni. These cities subsequently took grim places in U.S. battle history, because the Chinese offensive at these places inflicted heavy casualties on the American side. After the Chinese attack, Chinese forces retreated to the mountains. The U.S. air support did inflict heavy damage to the Chinese military, but the Chinese waited to continue attacking the U.S. armed divisions in massive numbers.

When Truman and General MacArthur met at Wake Island earlier that year, the general insisted to the president that China would not get involved if the United States invaded north of the 38th Parallel. This prediction by the general proved to be quite inaccurate, and this mistake deeply added to the mounting tension between Truman and MacArthur. After China did become officially involved in the Korean War, MacArthur lobbied the president for an all-out war between the United States and the recently adopted Communist nation of China, to which Truman said no.

From China's point of view, the new Communist country became involved in this conflict because it saw the U.S. intervention in the North Korean peninsula as being too close to the Chinese border. China reasoned that helping North Korea protect her land from aggression (America) was similar to the help that France provided to the United States during the Revolutionary War (pg 889, #4).

Truman himself wanted a limited war and was uncomfortable taking it to the level that MacArthur wanted. On November 26th, 1950, Chinese and North Korean forces attacked U.S. forces, slaughtering the 9th and 38th Divisions. The U.S. troops dealt with the invasion of 300,000 troops who came from the mountains. With this military engagement, MacArthur suffered his biggest defeat since Bataan in 1942. This defeat caused U.S.

forces to go on a 300-mile retreat, which the Marines described by the term "bug out."

By this time in late 1950/early 1951, the war began to take a serious toll on President Truman. At the Battle of Sunchon on December 1, 1951,, 7,000 U.S. soldiers were attacked, and 3,000 were either killed or wounded. Despite this terrible casualty/fatality list, the U.S. military did successfully defend Hungnam, where the Yalu River was located; 100,000 U.S. troops blew up the town, and left Hagaru-Ri in flames. U.S. air support was crucial for success in these endeavors.

Seoul was recaptured in early 1951 by the Communists. According to Colonel Joe Alexander of the Marine Corps, the Chinese tactics for fighting U.N. forces were to advance, attack, and withdraw. Because of this, the U.S. military figured out when to launch counter-offensives to specifically disrupt Chinese reorganization, which cost the Chinese Army many casualties. The *U.S.S. Missouri* played a role in this.

Even though Chinese leader Peng suggested a cease-fire to Chairman Mao Tse-tung, Tse-tung countered this proposal with a new attack against American/South Korean forces. This decision proved to be the worst mistake of the war for the Communists. This was so because the American forces had completely superior firepower to the Communists, and the U.S. Army was said to have toughened up through their rigorous combat experiences.

General Matthew Ridgway, who replaced General Walton Walker's 8th Division when Walker was killed in a car crash, had superior tank power compared to the Communists. General Ridgway launched "Operation Killer," which forced a Chinese retreat. Ridgway's forces were successful enough that by March 14th, 1951, the U.N. forces retook Seoul, and soon thereafter had reclaimed the land all the way to the 38th Parallel.

By April 5th, 1951, a letter was intercepted that was sent to Joe Martin from General MacArthur that revealed military information considered to be classified. This added to the already mounting tension that existed between Truman and MacArthur, because while the president began to seriously consider negotiating a peace treaty with both North Korea and China, MacArthur went on record that he would continue to fight the Chinese. MacArthur forgot a lesson that he learned at West Point: that the president, not the general, is the Supreme Military Commander. As a result, because Truman felt that world peace was more important, he felt obliged to fire General MacArthur, and did so on April 11th, 1951.

Though MacArthur felt publicly humiliated by President Truman, he came home to a ticker-tape parade and was truly celebrated as an American hero. MacArthur was invited to speak to a joint session of Congress, to which the general said, "Old soldiers never die, they just fade away." General Ridgway was Truman's replacement for MacArthur in Korea as the new Supreme Commander of U.N. forces.

By April 1951, the Korean War seemed never to go completely in favor of the U.N. forces. James Van Fleet, who was appointed leader of the 8th Army, had extensive military experience, and U.S. assistance to South Korea's army was steadily improving its strength. Soldiers from other countries, such as Britain, Greece, France, Australia, Canada, Turkey, Thailand, and the Philippines, proved to be reliable, dependable, and capable fighting soldiers. American B-29 bombers were used effectively against North Korean targets. The Soviet MiG-15s were the counterparts to the B-29. Due to superior American technology and equipment, the United States won the air fight versus North Korea ten to one.

Yet the Communists were still fighting strong. In the last week of April 1951, at the Injin River and Kapyong, a British brigade of 800 soldiers were brutally attacked over a five-day period. Only fifty of them survived. During this same time, the 8th Army was pushed back 35 miles and sustained 7,000 casualties. Incredibly, the 8th Army still managed to save the city of Seoul, and counter-attacked the Communist Army and inflicted 70,000 casualties. The Battle of Pork Chop Hill in 1953 was so heavy that it compared to World War One in its brutality.

Even after this apparent shift in momentum, the stalemate in the Korean War continued. The Communists were under pressure because the Soviets, who supplied the North Korean Army with weapons, wanted to be paid for them. While the negotiations for peace had begun, the talks had accomplished next to nothing. During this time, the Chinese military had launched a counter-offensive, which inflicted 60,000 casualties on the U.N. forces. The Communists themselves, however, suffered over 300,000 casualties.

Nevertheless, the Communists kept relocating their position. Old Baldy Hill was a vicious battle site of the Korean War. A military point system was set up by the U.S. military, so that if a soldier were to earn 36 points from bravery or service duty, he would earn the right to go home. In fact, the more dangerous the mission, the more points any one soldier could obtain. Many soldiers died in a vain attempt to fulfill this objective. Also, many American prisoners of war (P.O.W.s) in Korea were killed or beaten, even though North Korean P.O.W.s who were captured by the United States were freed.

At home, Truman's popularity was at its lowest point during his tenure. This factor, along with the overall toll that the presidency can bring to anybody, helped Truman decide not to seek the presidency in 1952. Truman even offered to fully support General Dwight Eisenhower for the presidency if Ike became a Democrat, a proposal Eisenhower promptly declined. Instead, Eisenhower ran for the presidency in 1952 as a Republican candidate. The Democrats chose to nominate Adlai Stevenson, who was then the governor of Illinois (pg 533, #3).

On election night, November 4th, 1952, General Dwight Eisenhower became the first Republican to obtain the White House in 24 years.

Eisenhower won the electoral vote for president with 442 votes, compared to Stevenson's 89 (pg 534, #3). General Eisenhower's service to his country during World War Two and Operation Overlord at Normandy Beach, France, earned him praise and admiration from the entire country. President Truman returned to his home in Missouri, where he lived until December 26th, 1972. He was 88 years old.

When Eisenhower became president, General Mark Clark succeeded General Ridgway, who then took over U.S. forces in Europe. Eisenhower promised during the 1952 campaign that if he was elected president, he would go to Korea, and he did. Eisenhower was elected president due to his heroism and proven competence as a general, statesman, and leader (pg 287, #5). In Eisenhower's policies dealing with the Korean conflict, he sought to find peace, because the thought of sending more boys to die, such as at the Battles of the Bulge and Normandy during World War Two, made Eisenhower see that the killing must stop (pg 331, #5).

President Eisenhower was successful in ending the Korean conflict by signing an armistice with North Korea, which ended the war and protected the South Korean position from the 38th Parallel to points south. General Clark was reluctant to sign the treaty, because the war's outcome would technically be ruled a tie. Nevertheless, General Clark signed the document at Panmunjom, Korea, in July 1953 (pg 538, #3). The U.S. troops came home, but some remained behind in order to protect the position of the 38th Parallel, where they remain today in defense of South Korea.

The U.S. Armed Forces, during their involvement in the Korean War from 1950-53, sustained 147,000 casualties and over 33,000 fatalities. The rest of the U.N. forces suffered over 400,000 dead or wounded, including South Korean fighters. The North Korean soldier deaths in the war were estimated to be in the hundreds of thousands. Yet the U.S. military, in unison with its South Korean and U.N. allies, achieved the victory of containing the spread of Communist aggression in the Korean peninsula, for which the United States, South Korea, and democracies worldwide can truly be grateful. Another war in another Asian province, almost a decade later, would lead our nation on a collision course with one of the most controversial and tragic wars that our nation ever fought: Vietnam.

Bibliography

1. History Channel videocassette:"The Korean War: Fire and Ice," Volume 1.
2. *Truman*, by David McCullough (previously listed).
3. *The Complete Book of U.S. Presidents*, by William Degregario (previously listed).
4. *The Wars of America*, by Robert Leckie (previously listed).

5. *Eisenhower: Soldier and President*, by Stephan Ambrose (previously listed).

Chapter 12
The Vietnam War (1959-73) and Presidents John Kennedy, Lyndon Johnson, and Richard Nixon

The Vietnam War, which lasted for fourteen years (1959-73), is described as an appalling human tragedy (pg xii, #1). The United States, France, and the Vietnamese people themselves would suffer an enormous amount of casualties in this prolonged conflict. Vietnam may have symbolically ended America's sense of Manifest Destiny (pg 9, #1). Millions of U.S. servicemen went to Vietnam, and over 58,000 died there. Their lives were given in a war where the U.S. military objectives were not met. This war involved the decisions of three U.S. Presidents: John Kennedy, Lyndon Johnson, and Richard Nixon, and their decisions would steer the course of the U.S. involvement in the conflict, until the eventual withdrawal of all military personnel in January 1973.

George Ball, who served in the State Department for both Presidents Kennedy and Johnson, proclaimed that the U.S. involvement in the Vietnam War was "probably the single greatest error made by America in its history" once the war was over. Former Defense Secretary Robert McNamara believed that America was damaged from the Vietnam War by the mistakes of the administrations involved regarding judgment and capabilities (pg xx, #2). McNamara and the U.S. military viewed the Vietnam conflict as a military operation, and not the "highly complex, nationalistic struggle" that it also was (pg 48, #2). In order to fully comprehend the Vietnam War and the presidential decisions of three leaders that led to the U.S. military involvement, it is important to understand the history of Vietnam as a whole, long before the U.S. combat battalions splashed ashore at Danang in March 1965.

Back in the year 208 B.C., a renegade Chinese general named Trieu Da conquered the territory of Au Lac, which is located in the northern mountains of Vietnam, and established a capital there. Da proclaimed himself "Emperor of Nam Viet" (pg 1, #1). Later in the first century B.C., the Han Dynasty expanded and incorporated Nam Viet into the Chinese Empire as the province of Giao Chi. Traders from as far back in history as ancient Rome used to travel to Vietnam to conduct business (pg 70, #1). The name of Giao Chi province would change over time.

Emperor Dinh Bo Linh ascended to the throne in 967 A.D. Linh called his state Dai Co Viet, and a period of independence followed. In 1428, the Chinese recognized Vietnam's independence by signing an accord, after nearly a decade of revolt that was led by Emperor Le Loi. Le Loi employed guerilla tactics to successfully repel the invading Chinese under the Ming Dynasty in 1426 (pg 116, #1). Even Mongol Emperor Kublai Khan invaded Vietnam in the 13th century, but was successfully repulsed by the Vietnamese.

Comprehensive legal codes and other reforms were implemented in Vietnam during the years 1460-98, when Le Thanh Tong was the ruler. However, by 1545, Vietnam was suffering from internal civil strife, which split the country for nearly two centuries. In 1627, a French missionary named Alexandre de Rhodes adapted the Vietnamese language with the Roman alphabet, which further paved the way for French influence in Vietnam. In fact, the French helped the Nguyen clan back to power after they had been ousted in the 1772 Tayson Rebellion. Also by 1787, Louis the XVI in France had helped Nguyen regain power.

However, by 1789, the French Revolution was underway, resulting in the deaths of Louis the XVI, Marie Antoinette, and many other people who were in power. In their wake would rise Napoleon, who would become consul for life in 1802. A Vietnamese leader named Gia Long (Nguyen Anh) unified the country in 1802 by becoming the emperor. A permanent deployment of the French Naval Fleet off the coast of Vietnam had been ordered by 1843. Consequently, the inevitable clash of French forces and Vietnamese mandarins occurred at he city of Tourane, now called Danang. Vietnamese leader Tu Dac ascended the throne that year with plans to eliminate Christian influence in Vietnam.

By 1852, Napoleon the III had taken power in France, and endorsed a series of expeditions to Vietnam in order to protect French missionaries stationed there and to gain concessions. During 1861, the same year that the Civil War exploded with a vengeance in America, French forces captured Saigon. The Vietnamese leader Tu Dac signed a treaty with the French, granting the French broad religious, economic, and political concessions. French control of the region extended into Cambodia in 1863.

Ten years later, French inroads into Tonkin began. French civilian governors in Cochinchina were appointed and extended their authority over Annam and Tonkin, in what France called a "protectorate." By 1887, France created the Indochinese Union of Annam, Tonkin, Cochinchina, and Cambodia. In 1890, the most influential leader in Vietnam's history to date was born in central Vietnam and was originally named Nguyen Sinh Cung. He would later change his name to Ho Chi Minh. In 1911, Minh left Vietnam, and he did not return to that land for 30 years (pg 5, #1).

In 1913, Minh moved to the United States. He was fortunate to be able to speak French, English, Russian, and three Chinese dialects. Minh relocated to Paris in 1918, and he remained there for the next seven years. Despite being away from his country, Minh would always hold Vietnam's interests very close to his heart. In fact, in 1919, Minh attempted to petition President Wilson for Vietnam's self-determination at the Versailles Peace Conference. Minh tried to ally himself with the United States to rid Vietnam of France. However, Minh was unsuccessful in getting U.S. cooperation to do so.

In December 1920, Minh joined the newly formed French Communist Party. Minh took his Communist beliefs a step further by moving from Paris to Moscow to become a full-time Communist agent. Minh assisted a Soviet representative in China named Mikhail Borodin, and he grew stronger ties with the Soviet Communist regime. Minh and his Communist comrades formed the Indochinese Party in Hong Kong in 1930. However, in Vietnam, it was the French who ran the region, and they had no problem showing their authority to the Vietnamese people.

A man named Bao Dai returned to Vietnam from a school in France to ascend the throne as emperor under French authority in 1926. The Popular Front Government in France sponsored a short-lived liberal reformation in 1936. World War Two was in full swing in 1941, so Minh returned to Vietnam covertly and formed the Vietminh to fight both Japan and France. These two nations posed the biggest threat to Minh's vision of Vietnam.

The Japanese attacked America on December 7th, 1941, at Pearl Harbor, which began America's involvement in the conflict. Japanese forces took over for the French administration throughout Indochina on March 9th, 1945. Bao Dai proclaimed the independence of Vietnam under Japanese auspices on March 11th, 1945. American President Franklin Roosevelt died on April 12th, 1945, succeeded by Harry Truman, the 33rd President of the United States.

By May 8th, 1945, Germany had formally surrendered. At the Potsdam Conference in July, Allied leaders assigned Britain to disarm the Japanese in southern Vietnam, and Chinese nationalists performed the same duty north of the 16th Parallel. After the United States dropped two atomic bombs on Nagasaki and Hiroshima, Japan officially surrendered on August 15th, 1945. The Japanese transferred power over Indochina to the Vietminh on August 18th, and Bao Dai abdicated his throne on August 23rd. Minh officially modeled the declaration of the independence of Vietnam after that of the United States. Minh was known to shelve his Communist principles from time to time in order to achieve his goals.

However, this independence was not meant to be, because British general Douglas Gracey and his military forces landed in Saigon on September 13th, 1945, and soon returned authority to the French government. Even President Truman believed that France should rule

Vietnam (pg 163, #1). However, in 1945, the situation was so bad in Vietnam that two million people—out of the total population of ten million—died of starvation (pg 166, #1). Once back in power, the French negotiated with Minh to grant the independence of Vietnam with French military assistance, but the French would later retract this proposal. France did this because it wanted full control of Vietnam.

Consequently, French military involvement against Vietnamese forces began in 1946 and lasted until 1952, when the French were defeated. In fact, by 1952, 90,000 French soldiers had died, been wounded, or been captured after six years of war (pg 203, #1). Despite this, President Truman heavily supported France financially throughout its war in Vietnam (pg 185, #1). These payments were authorized by President Truman and Secretary of State Dean Acheson (pg 192, #1). The U.S. government felt that it could not support Ho Chi Minh because of his endorsement by the Soviet Union. This is how France's military involvement in Vietnam began in 1946.

That year, China agreed to withdraw its forces from Vietnam, and France conceded its extraterritorial rights in China. France and the Vietminh reached an agreement in March 1946 in which France would recognize Vietnam as a "free state within the French Union" (pg 8, #2). French troops were then authorized to return to the north to replace the Chinese. A referendum was formed to determine whether Tonkin, Annam, and Cochinchina should be reunited. In May 1946, Ho Chi Minh went to Fontainebleau for negotiations, representing Vietnam.

The French high commissioner for Indochina, Admiral Georges Thierry d'Argenlieu, violated the March 1946 agreement by proclaiming a separate government for Cochinchina in June. The negotiation process between Minh and the French at Fontainebleau would break down in September. However, Minh signed a "modus vivendi" covering economic issues and agreed to a cessation of hostilities between France and Vietnam. As a result, Ho Chi Minh returned to Vietnam.

Nevertheless, French warships bombarded Haiphong, Vietnam on November 23rd, 1946. The Vietnamese Army retaliated by attacking French garrisons in Vietnam. Ho Chi Minh created a rural base, and the war in Vietnam officially began against the French. During 1947, President Truman and then-Secretary of State George Marshall outlined a plan to combat Communist insurgencies worldwide, a policy that would soon be called the Truman Doctrine. Accordingly, Congress approved a plan on May 15th that sent financial aid to the countries Greece and Turkey to help them fight off the Communists, who were attempting to take over their governments, and the plan was successful. Vietnamese leader Bao Dai, then living in Hong Kong, attempted to begin negotiations with France to achieve Vietnam's independence.

Dai signed an agreement with the French High Commissioner Émile Bollaert in December, which recognized Vietnamese independence from

French rule, but it contained limitations. Later, on March 8, 1949, Bao Dai and French President Vincent Auriol signed the Elysee Agreement, which made Vietnam an "associated State" within the French Union. This meant that France would retain control of Vietnam's defenses and finances. As a result, Bao Dai returned to Vietnam after three years of self-imposed exile.

During that same year, Chinese leader Chiang Kai-shek was driven from mainland China after a successful Communist revolution there, led by Mao Tse-tung, who proclaimed the establishment of the People's Republic of China by October 1st, 1949 (#2). Sensing the global Communist threat to free nations, the North Atlantic Treaty Organization (N.A.T.O.) was formed by the United States, Canada, and ten other western European nations on August 24th, 1949. The clash between democracy and communism seemed destined for a collision course.

Accordingly, Ho Chi Minh declared on January 24th, 1950 that the Democratic Republic of Vietnam is the only legal government. It was recognized by the Soviet Union and China. Yugoslavia also established diplomatic relations with Vietnam, which led some U.S. officials to suggest that Minh was not a "Soviet puppet." In response, both Britain and the United States recognized Bao Dai's government on February 7th, 1950.

American troops were called to action in Korea when the Communist north invaded the south of Korea by crossing the 38th Parallel. Almost simultaneously, President Truman signed legislation that granted the French military $15 million in aid for its war in Indochina on July 26th. Yet the French were defeated by the Vietcong at a key post on the Chinese border. When the Korean War concluded in 1953, the U.S. military successfully contained communism north of the 38th Parallel. That same year, France granted the country of Laos full independence as a member of the French Union in October.

The majority in the French National Assembly expressed hope for a negotiated settlement to the Indochina War. At the same time, the country of Cambodia declared its independence from France. It was led by Prince Norodom Sihanouk, who took control of the Cambodian Army on November 9th, 1953. France responded by reoccupying Dienbienphu. The Vietminh forces responded by pushing into Laos in December. Though Minh told a Swedish newspaper that he wanted to discuss peace proposals with France, his forces defeated the French at the Battle of Dienbienphu, which began on March 13th, 1954.

President Eisenhower decided against sending U.S. military intervention to Indochina to help France. Eisenhower did not commit U.S. troops to Vietnam because he had no mandate from Congress, and he had no military support from Great Britain (pg 213, #1). Meanwhile, in 1954, Bao Dai selected Ngo Dinh Diem to be the Prime Minister of Vietnam on June 16th. With the war going badly for the French, they began to negotiate with the Vietnamese, and agreements were reached in July. This agreement called for a cessation of hostilities in Vietnam, Cambodia, and Laos. The provisional demarcation line was drawn at the 17th Parallel that divided Vietnam, pending a political settlement to be achieved through nationwide elections.

The final declarations were accepted orally by all the participants at the conference, except for the United States, which stated that it had no intention to disturb the agreements but would view renewed aggression with concern. Bao Dai's government, on the other hand, denounced the agreements. The Southeast Asia Treaty Organization (SEATO) was created on September 8th , 1954, by a coalition of nations: Great Britain, France, Australia, New Zealand, Pakistan, Thailand, the Philippines, and the United States (#2). Though President Eisenhower did not send military support to the Vietnamese leader Diem, he did send his special envoy, General J. Lawton Collins, to Saigon to affirm U.S. support for Diem. This agreement included $100 million in aid.

The U.S. Navy helped hundreds of thousands of refugees from North Vietnam flee to the south. As previously stated, the U.S. government believed that it was impossible to support Ho Chi Minh because he was a Communist, and was also endorsed by the Soviet Union (pg 194, #1). However, President Eisenhower stated that he would support Diem militarily if he showed that he was capable of leading a free Vietnam (pg 235, #1). Accordingly, in January 1955, the U.S. began to funnel aid directly to the Saigon government, and also agreed to train the South Vietnamese Army. Consequently, Diem went on to crush the Binh Xuyen military sect in April 1955.

Diem rejected the Geneva Accords and refused to participate in nationwide elections on July 16th, a decision backed by the United States. Diem defeated Bao Dai in a referendum on October 23rd, 1955, and Diem would become Vietnam's Chief of State. Diem proclaimed the existence of the Republic of Vietnam, with himself as President, on October 26th, 1955. Meanwhile, Ho Chi Minh was in Moscow in July 1955, and he accepted aid from the Soviet Union. He successfully negotiated for Chinese assistance in Beijing, also.

In December 1955, land reform in North Vietnam reached its most radical phase, as landlords had to go before the "people's tribunals." By January 1957, the Soviet Union favored a permanent division of North and South Vietnam, and proposed that the two territories enter the Untied Nations as separate nations. President Eisenhower proclaimed his support for the Diem regime and invited Diem to visit the United States in May for a ten-day visit, which Diem accepted. The Communist insurgency in the north decided, in Hanoi, to organize 37 armed companies in the Mekong Delta. During 1957, Communist guerillas assassinated more than 400 minor South Vietnamese officials. The Communists formed a coordinated command structure in the Eastern Mekong Delta in June 1958.

By 1959, North Vietnam formed Group 559, whose cadres began transporting weapons into South Vietnam via the Ho Chi Minh Trail. By July 1959, Group 759 was organized to send supplies into the south by sea. U.S. Commanders Major Dale Buris and Sergeant Chester Ovnand were killed by Communist guerillas at Bienhoa on July 8th, 1959, which marked the first American deaths in what would officially be called the Vietnam era.

Diem promulgated the law that authorized intense repression of Communist suspects and other dissidents in August 1959. Hanoi leadership responded by creating Group 959 in September 1959, in order to furnish weapons and other supplies to Communist insurgents in Laos. By 1960, the North Vietnamese imposed a universal military conscription in April. Diem was petitioned by his people to reform his government. A Workers' Party leader named Lao Dong opened Congress in Hanoi on September 5th, 1960, and he stressed his desire for people to combat the Diem regime.

In fact, the South Vietnamese Army unsuccessfully attempted to overthrow Diem on November 11th, 1960. Also in 1960, leaders in Hanoi organized the National Liberation Front for South Vietnam on December 29th, which the Saigon regime dubbed the Vietcong. This meant the Communist Vietnamese. The Vietnamese people sided with whichever government they felt would harass them the least, and this equation favored Minh over Diem. Meanwhile, back in America, Senator John Kennedy won the presidency over Vice President Richard Nixon on November 8th, 1960. It was during the Kennedy Administration that the U.S. policy on Vietnam changed to further combat the spread of global communism.

When Kennedy was inaugurated as president, outgoing President Eisenhower warned him that the country of Laos would be the major crisis in Southeast Asia. Yet in Vietnam, under the Diem regime, the Vietnamese people were very poor and malnourished (pg 249, #1). The supposed democratic election in Vietnam that catapulted Diem to power in 1960, with America's help, was a complete sham. Diem's brother Nhu also had power in Vietnam. The United States pumped more than $7 billion into Vietnam's economy from 1955 to 1961 (pg 31, #2). President Kennedy was in a full-blown international crisis over the spread of communism internationally and had taken some heat in April 1961 for the failed invasion, by 1,500 Cuban expatriates, to topple Fidel Castro's Communist regime in Cuba; they did not receive the proper U.S. military support that was promised to them. After three days of fighting, 1,100 of the survivors surrendered and were captured by the Communists (pg 555, #3).

In order to get the prisoners released, the United States paid Cuba $53 million in food and medical supplies, and they were released in 1962 (pg 555, #3).The Eisenhower Administration had originally planned the Bay of Pigs invasion, but it was Kennedy who approved the military plans to go forward, and he ultimately took all the blame for failing to achieve these goals. In fact, the invasion itself was a complete failure. As a result, President Kennedy decided that all future military decisions would come from Kennedy and his own men, and not from the Joint Chiefs or intelligence agencies. Kennedy still wanted to appear to the global community as being tough on communism, and was determined to prevent its influence from spreading into other foreign countries.

Kennedy's meeting with Soviet Prime Minister Nikita Khrushchev in Vienna on June 4th, 1961, was tense and potentially dogmatic, as the two leaders clashed over the core ideas and philosophies of the two countries. For example, the Soviet Union was successful in launching the first manned spacecraft, ahead of the United States (pg 106, #4). Kennedy felt the pressure to keep up with the Soviet Union in the race to outer space. In fact, it was Kennedy who wanted to surpass the Soviets in outer space by landing an American on the moon and returning him safely to the Earth by the end of the 1960s. As a result, dramatic federal funding for NASA increased.

Nevertheless, Kennedy's mission in the Vienna meeting with Khrushchev was meant to "preserve the status quo" between the two nations (pg 157, #4). However, Khrushchev was very rough on Kennedy and treated him like a little boy. After all, Khrushchev's son was older than Kennedy. Khrushchev told Kennedy that 20 million Russians died in World War Two at the hands of the Nazis, and, therefore, he did not want to see a reunited Germany (pg 168, #4). The Soviets also wanted to see the United States leave Berlin, to which President Kennedy replied no. In fact, during this meeting with Khrushchev in Vienna, Kennedy feared that nuclear war with the Soviet Union was inevitable (pg 550, #12). Kennedy was especially concerned about the situation in Berlin (pg 553, #12).

In Ted Sorensen's biography on President Kennedy, he stated that Kennedy observed that "America's role as world leader often involved it in disputes between its friends and allies" (pg 580, #12). Sorensen would go on to explain that President Kennedy would regard a catastrophic nuclear war between America and the Soviets as the "ultimate failure," because it would have represented failures of deterrents, diplomacy, and reason (pg 625, #12). As much as Kennedy wanted to avoid war with the Soviets, he knew that the situation in Berlin could easily cause nuclear war between the two nations, because if the Soviets attacked Western Europe in any way, he would be forced to use nuclear weapons.

Kennedy believed that the place to confront the Soviets was in the country of Vietnam. Democratic governments' security, especially in the battle against global communism, depended in part on the United States maintaining a presence in Asia (pg 649, #12). Consequently, the American military assistance command was established by February 6th, 1962. By the middle of 1962, the American military adviser presence increased from 700 to 12,000. Kennedy sent these U.S. military advisers to train the South Vietnamese military (pg 653, #12). However, Kennedy was committed to keeping the conflict limited and avoiding war in Vietnam if possible (pg 653, #12). This policy was known as "containment." Kennedy feared that the Vietnamese people would turn against the United States, as they did with the French (pg 654, #12).

At around the same time, in May 1962, the Communists formed battalion-size units in central Vietnam. By this time, North Vietnamese troops outnumbered South Vietnam's troops ten to one (pg 653, #12). On July 23rd, 1962, the second Geneva conference was held, with representatives from the United States and 13 other nations. There, it was agreed that the independence and neutrality of the country of Laos should be guaranteed. This idea fueled Kennedy's beliefs that if Vietnam were free, then the countries of Laos, Cambodia, and Thailand would be free also (pg 182, #3). By mid-1962, Kennedy and his administration wanted to train the South Vietnamese army to fight for themselves against the north, and that would be the limit of U.S. participation. Despite this ideology, Kennedy raised the U.S. troop level in Vietnam to 15,500 by the end of 1963 (pg 654, #12).

Even though there was a U.S. troop increase in Vietnam in 1963, Kennedy said that South Vietnamese leader Diem had to inspire confidence and morale in the South Vietnamese troops, and that this was not America's responsibility (pg 654, #12). Though Kennedy knew that Diem was a controversial figure in governing South Vietnam, he chose to assist him anyway, with military advisers and equipment. This was consistent with Kennedy's Cuban Missile Crisis Speech in 1962, in which he said, "The 1930s taught us a clear lesson: aggressive conduct, if allowed to go unchecked and unchallenged, ultimately leads to war."

Vice President Lyndon Johnson said in 1961, on a trip to Vietnam, that Diem was the reincarnation of Winston Churchill. While Kennedy did not want to send troops to Vietnam, he did not want to accept the defeat of the South Vietnamese Army, either (pg 268, #3). Kennedy believed that if Vietnam was taken over by the Communists, so too would go all of Southeast Asia.

When the Berlin Wall was erected to stop the flow of refugees from East Germany to the west in 1961, Kennedy was relieved that he did not have to go to war with the Soviets. While Kennedy was criticized at home for his inaction to stop the construction of the wall, he responded by saying, "It is better to have a wall than a war" (pg 246, #3). However, Soviet Premier Khrushchev began to detonate nuclear bombs in Siberia to prove that he was catching up to the American military (pg 251, #3). Kennedy responded to the Soviet propaganda by conducting nuclear tests of his own. The grim estimates of the results of a nuclear war with the Soviets were approximately 70 to 80 million people dead in the United States, which was about half of the nation's population at the time. Just one enemy missile could kill up to 600,000 people, which was approximately the total amount of fatalities during the American Civil War.

Accordingly, Kennedy called for an additional $6 billion for defense spending and raising more personnel. From October 16th to 28th, 1962, democracy and communism came very close to global destruction when U.S. spy satellites uncovered Soviet offensive nuclear missile bases in Cuba that were capable of striking the eastern two thirds of the United States, as well as Latin America. These spy satellites showed that the Soviets were constructing offensive nuclear bases in Cuba, contrary to the Soviet claim that they were merely defensive missiles. Also, the Soviet troop presence on the island of Cuba was reported to be around 40,000 (pg 345, #3).

The Kennedy Administration was already estimating the numbers of a U.S.-led invasion of Cuba, which turned out to be very high. Kennedy's other options were limited air strikes and a naval blockade. Uncertain of the decision he had to make, Kennedy told his speechwriter, Ted Sorensen, to write two different speeches for him to deliver to the American people: one for the naval blockade and the other for the actual bombing of Cuba (pg 383, #3). In order to do this, Sorensen researched declarations of war by Presidents Woodrow Wilson and Franklin Roosevelt in 1917 and 1941, respectively. Kennedy's Cabinet men McGeorge Bundy, Dean Acheson, and General Max Taylor urged the president to use air strikes against Cuba, and claimed that time was running out to effectively do so (pg 384, #3).

Kennedy debated the idea of asking Congress for a formal Declaration of War against the Soviet Union. Dean Acheson, who favored air strikes, considered Khrushchev to be a "madman." Attorney General Robert Kennedy, contrary to most of Kennedy's Cabinet, favored a naval blockade at first, then an air strike on Cuba, but only if necessary. The Kennedy Administration actually considered giving up U.S. missile bases in Turkey and Italy to diffuse the situation (pg 392, #3). Kennedy faced the possibility of millions of people being killed within hours in the event of a nuclear war with the Soviet Union (pg 393, #3).

On October 22nd, 1962, President Kennedy went on television to give a speech to the American people, and to the people of the world, announcing that there would be a naval blockade around the country of Cuba to stop Soviet ships from sending nuclear material to the island. Air strikes against the country of Cuba were set to begin at dawn on October 30th, 1962, under an Executive Order. On October 27th, however, Khrushchev backed down and agreed to remove all nuclear material and missile sites on Cuba.

Khrushchev then offered a nuclear test ban treaty that would stop all testing (pg 439, #3). This was a clear Kennedy victory in the Cold War. But Kennedy's problems with communism were just beginning in Vietnam. Because of charges of corruption in Ngo Dinh Diem's administration, Kennedy questioned Diem's ability to lead a democratic South Vietnam (pg 442, #3). In fact, President Kennedy was hearing many contradictions about the progress of the Vietnam War (pg 443, #3). Kennedy had increased the U.S. military presence in Vietnam to 11,500 personnel by the end of 1962. That number had increased by 9,000 from just the previous year. The number of fatalities among American troops had also increased, from 14 at the end of 1961 to 109 by the end of 1962.

While the Defense Department was more optimistic about American chances to win the Vietnam campaign, the State Department was more pessimistic. Yet both the Defense and State Departments blamed Diem for the lack of victory (pg 444, #3). Conflicting American military intelligence reports came in that the Vietcong military had lost 20,000 men who perished in battle, and that an additional 4,000 had been wounded. However, the Vietcong still managed to increase their numbers from 18,000 to 23,000 (pg 447, #3). Though the United States believed that it was winning the war in Vietnam, it was going a lot slower than anticipated.

Kennedy believed that the threat to the Vietnamese people was a threat to the free world. In fact, President Kennedy believed that the fall of South Vietnam to the Communists meant that communism would soon spread into the Western world, even to the United States, if left unchecked (pg 29, #2). Yet Diem himself was almost assassinated in November 1962 by the South Vietnamese (pg 281, #1). By 1963, the Vietnam conflict escalated into a different, more intense war, based on a series of historical events in both Vietnam and America. On January 2nd, 1963, the South Vietnamese Army was defeated by the Vietcong at the Battle of Ap Bac. To compound the problem, Diem and his South Vietnamese government shot at Buddhist demonstrators in Hue on May 8th. In June, a Buddhist monk set himself on fire to protest Diem's policies. In fact, on July 4th, 1963, South Vietnamese General Tran Van Don informed C.I.A. agent Lucien Conein that the south's army was plotting to overthrow Diem.

Nevertheless, Kennedy's military advisers believed that the United States could win the Vietnam War either by the end of 1963 or in just two or three years (pg 482, #3). In fact, the plan was to pull out 1,000 U.S. advisers from Vietnam by the end of 1963 (pg 49, #2). The overall Kennedy Administration plan was to have all American troops withdrawn from Vietnam by the end of 1965 (pg 78, #2). Kennedy authorized the U.S. Air Force to use napalm on the North Vietnamese. After some military successes by the south, the South Vietnamese government informed the United States that its troop numbers could be reduced to 12,000 or 13,000. Based on this information, Kennedy believed that some, if not all,

U.S. troops in Vietnam could start coming home by the end of 1963. Yet Kennedy was understandably ambivalent about this timetable.

Kennedy sent Ambassador Henry Cabot Lodge to Saigon on August 22nd, 1963, to analyze the situation and the stability of South Vietnam. After all, Kennedy and Soviet Premier Khrushchev signed a nuclear test ban treaty on July 25th, 1963, and Kennedy did not want to let his opportunity to save South Vietnam slip away. However, Lodge's report to Kennedy about the South Vietnamese situation was grim, and it was recommended that both Diem and his brother Nhu be overthrown. The Kennedy Administration approved this recommendation. On August 24th, 1963, Lodge received the order from President Kennedy that Diem should be removed from power in Vietnam (pg 55, #2 and pg 568, #3). Kennedy sent Secretary of Defense McNamara to Saigon on September 23rd, 1963, to appraise the situation and to meet with Diem twice (pg 69 #2). However, this proposal was adopted because Kennedy's Cabinet members knew that the Vietnam War could not be won with Diem and Nhu in charge.

Consequently, President Kennedy gave Ambassador Lodge the power to suspend all U.S. aid to Diem in the hopes that he would be overthrown. The United States did not want to lose Vietnam to the Communists as they did China in 1949 (pg 604, #3). Kennedy would go on to criticize Diem in a television interview on September 2nd, 1963 with CBS's Walter Cronkite, in which he said that the South Vietnamese Army had to win or lose the war, and America was only there to advise (pg 62, #2). Kennedy believed that if the South Vietnamese could not defend themselves on their own, even with American military advisers, then the war in Vietnam could not be won. Yet Kennedy also said that an early withdrawal from Vietnam would be a "great mistake." Diem knew that on November 1st, 1963, there would be a coup launched against him and his brother. Diem was plotting a counter-coup of his own to crush the rebels, which was led by his brother Nhu. Unfortunately for Diem and Nhu, events in the uprising went terribly wrong for the two brothers (pg 321, #1).

After their forces were overwhelmed in battle, both Diem and Nhu had the opportunity to go quietly into exile, which both refused. The United States refused to grant asylum to the brothers. Eventually, on the run and out of power, the two brothers agreed to surrender on November 2nd, 1963. Unfortunately for them, the two brothers were murdered that day in a car while attempting to escape the country. This ended the Diem/Nhu regime in South Vietnam (pg 326, #1). Tragedy would also strike America with a vengeance three weeks later, when President Kennedy was murdered in Dallas, Texas on November 22nd, 1963, by a Communist sympathizer. This single act may have altered the course of U.S. involvement in the Vietnam conflict; the U.S. presence there might have decreased had Kennedy and Diem not been assassinated. In fact, speculating after the assassination of President Kennedy, Bob McNamara wrote in his book, *In Retrospect*, that had President Kennedy lived, he would have pulled all U.S. advisers out of Vietnam.

Kennedy was succeeded by Vice President Lyndon Johnson. By the end of 1963, 15,000 American military advisers were in South Vietnam, and the South Vietnamese government was receiving $450 million in financial aid from America. In the first nine months of President Johnson's Administration, six South Vietnamese governments had come and gone (pg 101, #2). In response, the Communist leadership in Hanoi decided to step up the struggle in the south of Vietnam. General Minh replaced Diem in power for South Vietnam. Among the U.S. military personnel stationed in Vietnam, as an advisor in December 1962, was then-Captain Colin Powell, who would later rise to become a four-star general, Chairman of the Joint Chiefs, National Security Adviser to the President, and Secretary of State for President George W. Bush. This was the first of two tours of duty Powell completed for the U.S. Army in Vietnam.

In January 1964, General Minh, who took over the leadership of the south after the assassination of Diem, was overthrown himself by General Nguyen Khanh. Vietnam now seemed like a political vacuum with Diem dead. While Khanh was in command of South Vietnam, he allowed Minh to remain as the figurehead Chief of State of South Vietnam. Back in America, Lyndon Johnson took over the presidency in a wounded nation grieving from the assassination of President Kennedy. Once President Johnson assumed power, he was afraid of looking like a coward and Communist sympathizer by backing out of Vietnam (pg 336, #1). Johnson remembered his days in the U.S. Senate, when the McCarthy hearings were taking place, in the 1950s. Johnson did not want to be perceived by the public as being soft on communism. This happened to President Truman when China became a Communist nation in 1949. Therefore, Johnson wanted to pass a military resolution, unlike what Truman had done in the Korean conflict, that said if he had to increase troop and fighting levels in Vietnam, then he could do so; he felt that it was a necessary function to have (pg 120, #2). Johnson feared that the international community would not let him accomplish his goals as president if he withdrew from Vietnam. Besides, Johnson thought that he could deal with Ho Chi Minh just like a businessman and make him political offers that he believed Minh could not refuse (pg 338, #1).

In order to understand President Johnson better, and the decisions that he would go on to make in regards to the U.S. military's involvement in the Vietnam War, it is important to study the history of the man from the very beginning. Lyndon Baines Johnson was born on August 27th, 1908 (pg 6, #5). Both his father, Sam Ealy Johnson, Jr., a farmer and a Texas politician, and his mother, Rebekah, graduated from Baylor University. Rebekah majored in journalism and went on to edit for a small local Texas newspaper (pg 564, #9). Sam Johnson lost his fortune when Lyndon was a boy, because he mortgaged his house on the success of his cotton production, which failed (pg 14, #5). These losses by his father reinforced Lyndon's sympathy for the needy and his sense of compassion for the farm worker.

Johnson was born in Stonewall, Texas, and went to high school in Johnson City, Texas, where he graduated in 1924. Johnson did not go to college when the opportunity first presented itself. Instead, Lyndon migrated to California for farm work. He would end up working for a man named Tom Marti, who was a lawyer in San Bernardino. This mentorship influenced Johnson to pursue a career in law. Before that, however, Johnson was a store clerk, fruit picker, dishwasher, and elevator operator (pg 565, #9).

Johnson was influenced by politics in his youth by his father, who served in the Texas House of Representatives. In fact, Johnson was deeply influenced by a visit with his father to the Alamo. After a string of jobs out west before college, Johnson returned to Texas flat broke in 1926. He went on to college at Southwest Texas State Teachers College, from 1927 to 1930. When Franklin Roosevelt became president in 1933, the New Deal programs that he implemented gave Johnson a belief that government can intervene to help the U.S. population in fighting the Depression and unemployment. This would become his political philosophy (pg 42, #4).

Lyndon married Claudia Alta "Lady Bird" Taylor on November 17th, 1934, while acting as a secretary on Capitol Hill for Democratic Representative Richard M. Kleberg (pg 567, #9). During this time, Johnson learned a lot about how the House of Representatives operated on a daily basis (pg 76, #6). Johnson briefly decided to attend Georgetown Law School in 1934, but dropped out shortly thereafter. Yet Johnson's hard work on Capitol Hill would pay off when, in 1935, President Roosevelt appointed him as the director of the National Youth Administration of Texas, whose mission was to provide jobs to thousands of young people suffering from the effects of the Great Depression (pg 84, #6).

Johnson's work with the National Youth Administration in Texas was a tremendous success. Opportunity arose anew for Johnson in 1937, when Representative James Buchanan (D) of Texas passed away on February 23rd. This event required a special election run-off to fill his Congressional seat. Johnson saw this as his opportunity to enter Congress. In order to win the election for Congress, Johnson had to completely back President Roosevelt and his policies. For example, posters would say, "A vote for Johnson is a vote for Roosevelt's program" (pg 61, #5). Johnson also heavily lobbied for the African-American vote. In the end, Johnson would go on to win the election, representing Texas's Tenth District. He was sworn into office on May 13th, 1937, at the age of 28. Johnson would go on to serve six terms in the House, until 1949 (pg 68, #5).

Johnson was reported to be having an affair with Alice Glass, which was rumored to have begun in the late 1930s. When World War Two began for America in December 1941, Johnson served as a lieutenant commander of the Navy from December 1941 to July 1942. He did this while he was still a member of Congress. In fact, Johnson's plane survived a Japanese aircraft attack while on an observation mission over New Guinea, for which Johnson was awarded a Silver Star (pg 567, #9). Johnson returned to Congress in July 1942 when President Roosevelt ordered all Congressmen to resume their duties in the House. Johnson served on the Naval Affairs Committee.

Johnson was defeated on his first attempt for a U.S. Senate seat from Texas in 1941, but would go on to win a Senate election in 1948. Johnson was elected minority whip in the Senate in 1951, and when the Democrats won control of both houses of Congress in 1954, Johnson then became majority leader. Johnson championed civil rights for minorities in a hostile south during this time (pg 147, #5). However, due to President Eisenhower's popularity, Johnson chose not to openly fight the president on political issues (pg 165, #5).

Johnson would introduce the "Johnson Doctrine," which called for the eradication of Communist influence (pg 170, #5). This bill passed sixty-nine to one. The causes that Senate Majority Leader Johnson was passionate about were education, public works, and the minimum wage. Johnson was chairman of a Senate subcommittee on aeronautical and space science, and he was very influential in its development. This is why the Johnson Space Center in Houston is called what it is.

In the presidential campaign of 1960, though, Johnson did not win the nomination over Senator Kennedy. But Kennedy asked him to be the vice-presidential candidate, due to Johnson's strength in the south and in the Senate. The ticket went on to victory in November and thrust Johnson from Senate majority leader to vice president. Yet Johnson did not have a lot of influence on Kennedy's Administration, largely traveling abroad to promote the president's interests. In fact, as vice president, Johnson visited Southeast Asia in 1961 to discuss the Vietnam situation with then pro-American President Ngo Dinh Diem (pg 271, #5). Johnson believed at the time that direct U.S. military intervention was not necessary. When Johnson was thrust into the presidency after the assassination of John Kennedy, his decisions on foreign policy against Communist aggression would become American policy, including Vietnam. Johnson, not Kennedy, had to campaign against Republican Senator Barry Goldwater in the 1964 presidential race, and he had to remain strong with his domestic policy goals, such as attacking poverty and promoting civil rights. Johnson's campaign slogan for the 1964 presidential election was the "Great Society" (pg 316, #5). Therefore, President Johnson did not immediately push the war's escalation in 1964, because he was consumed with his civil rights legislation (pg 122, #2).

Though Vietnam was not yet the problem it would eventually become, Johnson had to deal militarily with North Vietnam when they advanced past the 17th Parallel, the demarcation established by the Geneva Accords of 1954 (pg 317, #5). When this happened, Johnson decided to pursue the war because he felt that Kennedy would have done the same thing if he were president (pg 319, #5). Johnson sent Defense Secretary Bob McNamara to Vietnam in March 1964, where he vowed his support for General Khanh. Johnson even had McNamara take many pictures of himself and Khanh together (pg 112, #2). On June 2nd, McNamara, Dean Rusk, and other Cabinet officials met in Honolulu, Hawaii, and they agreed to increase aid to South Vietnam. The Pentagon strategists began to refine their plans for bombing North Vietnam with American airpower. In fact, covert South Vietnamese maritime operations began against the north in July 1964.

The U.S. military situation dramatically escalated when North Vietnamese patrol boats attacked the U.S. vessels *Maddox* and *Turner Joy* on August 2nd, 1964, which gave Johnson extraordinary power to act militarily in Southeast Asia. Johnson addressed the nation on television that evening by saying, "Aggression by terror against the peaceful villages of South Vietnam has now been joined by aggression on the high seas against the United States of America" (pg 321, #5). However, Johnson sought "no wider war" at the time.

The Vietcong Army attacked the Bien Hoa air base on November 1, 1964, and succeeded in destroying six B-57s and 20 additional aircraft. This resulted in five U.S. soldiers killed and 100 injured (pg 418, #1). Despite this, Johnson rejected proposals from his advisers for retaliatory raids against North Vietnam. Meanwhile, back in the United States, the 1964 presidential election was about to happen. Senator Barry Goldwater, Johnson's opponent, voted against the civil rights bill and the Economic Opportunity Act (pg 329, #5). Johnson was heavily in favor of both of these issues, as well as the later Voting Rights Act, federal legislation that forced all states to guarantee African-Americans the right to vote (pg 359, #5).

On election night, November 3rd, 1964, Johnson defeated Barry Goldwater by 61% to 39%, or 486 electoral votes to Goldwater's 52 (pg 571, #9). Even though Johnson beat Goldwater, he felt the pressure of his opponent calling him soft on communism. When Johnson was elected president in November 1964, he strongly pushed his Great Society programs through Congress, in which he would offer as much as he could to the poor and disenfranchised. He based his legislative programs on Franklin Roosevelt's New Deal ideas, such as programs producing food and jobs and improving the lives of African-Americans, women, and Native Americans.

Before 1964 was finished, Saigon was overwhelmed by rioters, who were protesting the rule of General Nguyen Khanh. As a result, U.S.

Ambassador Maxwell Taylor urged Khanh to leave the country. Meanwhile, Vietcong terrorists bombed the U.S. military post located in Saigon on Christmas Eve. The newly elected president did not want to appear soft on communism, and would often act as his own chief of staff. When Johnson was unfamiliar with foreign policy, he would rely on his Cabinet to influence him in his decisions (pg 256, #5). As president, Johnson thought that the United States would look vulnerable and weak in the world's eyes if it did not stop the Communists in north Vietnam from invading the south.

Johnson sent his National Security Adviser, McGeorge Bundy, to Saigon on February 4th, 1965, when Soviet Premier Alexei Kosygin was in Hanoi. Two days later, on February 6th, 1965, the Vietcong attacked the U.S. military advisers' barracks in Pleiku, which resulted in nine U.S. soldiers being killed. After the Gulf of Tonkin Resolution was passed by both houses of Congress, it became the guiding legislative document by which the United States participated in the Vietnam War (pg 575, #9). Johnson realized that in order to win the war, he would have to "Americanize" it and get it out of the hands of the incompetent South Vietnamese officials.

Accordingly, after the attacks at Pleiku, Johnson wanted to strike back. Bundy recommended to Johnson a sustained and graduated bombing of North Vietnam, which began on February 24th, 1965. He ordered U.S. air raids to take place against North Vietnam, officially called Operation Flaming Dart. He had the authority to do this from the passage of his Gulf of Tonkin Resolution in Congress (pg 171, #2). Johnson also wanted to send $1 billion in economic aid to both Vietnam provinces in order to achieve economic and social betterment (pg 266, #5). Meanwhile, back in Vietnam, General Khanh left the country, and Dr. Phan Huy Quát formed a new government in Saigon on February 18th, 1965. The sustained campaign of American bombing of the north began on March 2nd, 1965, and was given the name "Operation Rolling Thunder". By March 8th, two U.S. Marine battalions, consisting of 3,500 men, landed in Vietnam to defend the Danang airfield. These were the first U.S. combat troops in Vietnam.

By the end of April 1965, 50,000 more U.S. troops arrived in Vietnam. Yet again, Johnson did this via the power derived from the Gulf of Tonkin Resolution, so he did not have to acquiesce to Congress in order to make military decisions (pg 275, #6). However, on April 7th, 1965, Johnson offered Ho Chi Minh participation in a Southeast Asian development program in exchange for peace while giving a speech at Johns Hopkins University. Unfortunately, North Vietnamese Prime Minister Pham Van Dong rejected Johnson's entreaty the next day, because Dong insisted that the settlement of Vietnam must be based on the Vietcong program.

On June 11th, 1965, Air Vice Marshall Nguyen Cao Ky took over as prime minster of the military regime in Saigon. The U.S. command in

Saigon reported on June 26th that the Vietcong military was responsible for placing five South Vietnamese combat regiments and nine battalions out of action. Sensing that the South Vietnamese were incapable of fending off the Vietcong Army on their own, Johnson placed 18 U.S. combat divisions in Vietnam. On July 27th, 1965, both Senators Mike Mansfield and Richard Russell tried to talk Johnson into leaving Vietnam, but Johnson disagreed and chose to send more personnel. Even Vice President Hubert Humphrey opposed the Vietnam War, and he was subject to Johnson's wrath because of it (pg 428, #1). But by mid-1965, it was clear to President Johnson that the only way to prevent a Communist takeover of South Vietnam was massive U.S. military intervention, which is what he decided to implement (pg 188, #2).

By July 28th, 1965, Johnson approved General William Westmoreland's request for an additional 44 combat battalions. The U.S. Congress could have stopped increasing troop activity in Vietnam without Johnson's approval, but it did not (pg 324, #6). In fact, the majority in Congress, at the time, did support the war. President Johnson cited the examples of President Wilson not reacting against the aggression of the Kaiser until it was too late, and President Franklin Roosevelt not dealing with Germany/Japan quickly enough to have prevented World War Two. Johnson ultimately decided that, as the leaders of South Vietnam had proved so inept since the death of Diem, the only way to properly solve the conflict was with heavy U.S. military involvement. Johnson was determined not to be the first U.S. president to lose a war (pg 350, #1). These reasons provided self-justification to Johnson that the American involvement in the Vietnam conflict was just and warranted (pg 329, #6).

In September 1965, Chinese Defense Minster Lin Biao indicated to the United States that China would not directly intervene in the Vietnam conflict. Their leader, Chairman Mao Tse-tung, began the Great Proletarian Cultural Revolution. Meanwhile, back in Vietnam, American forces defeated Vietcong units at the Ia Drang Valley in November. This battle, near the Cambodian border, from November 14th to 19th, 1965, was a U.S. military victory, with 1,300 North Vietnamese soldiers killed and only 300 U.S. soldier fatalities. This was the first big conventional clash of the war. By the end of 1965, there were 185,000 U.S. troops stationed in Vietnam. The U.S. had already sustained 1,350 fatalities as a result. Also, the north could move more than 200 tons of supplies down the Ho Chi Minh Trail every day, despite the intense bombing campaign from the United States.

On December 25th, 1965, President Johnson suspended the bombing of North Vietnam in an attempt to negotiate with the Communists. When those attempts failed, Johnson resumed the bombings on January 31st, 1966. One of the advantages of the Vietcong was that it had the ability to recruit forces from South Vietnamese sympathizers (pg 236, #2). Johnson formulated a four-phase strategy to win the Vietnam conflict:

1.	Deploy troops to protect American boats along the coast and around Saigon

2.	Send forces to the central highlands to block any Communist attempts to cut the country in two

3.	Launch search-and-destroy missions to entice the enemy into battle, where U.S. firepower would decimate Communist forces

4.	Install a mop-up operation by bombing the Communists to choke off the supply flow, while also promoting pacification programs to try to win the hearts and minds of the Vietnam villagers (pg 384, #5)

These tactics, ultimately, did not work. Yet Johnson met with South Vietnamese leaders in Honolulu, Hawaii, on February 8th, 1966, to emphasize the need for the pacification of South Vietnam. However, South Vietnam had enough problems of its own: Buddhist demonstrators marched against the Saigon regime at Hue and Danang on May 23rd, 1966. As a result, South Vietnamese government troops took over Danang on May 23rd, and also Hue on June 16th, 1966. President Johnson had angrily disagreed with this decision when it was presented to him by his military advisers. Johnson also strongly disagreed with increasing U.S. troop numbers in Vietnam by an additional 113,000, because the president accused these military leaders of trying to provoke World War Three (pg 385, #5). Johnson wanted to win the Vietnam War, but he was uncertain about how to properly solve the conflict (pg 147, #2).

This concern was amplified by French President Charles de Gaulle's visit to Cambodia in September 1966, where he publicly called for American military withdrawal from Vietnam. In mid-1966, the Joint Chiefs were calling for 500,000 U.S. troops to be serving in Vietnam in 18 months' time. But by the end of 1966, the American troop numbers in Vietnam reached almost 400,000. It was Secretary of Defense McNamara who requested that President Johnson increase U.S. troop presence in Vietnam from 179,000 to 368,000 by the end of 1966. In response, the North Vietnamese increased their military size, also (pg 236, #2).

On January 28th, 1967, the North Vietnamese Foreign Minister Nguyen Trinh informed America that the United States must stop bombing North Vietnam before negotiation talks could begin. The United States felt that it had superior firepower in the war and that victory would soon be achieved. However, the North Vietnamese Army was being supplied with the latest Soviet weaponry, which gave it a military advantage, combined with its knowledge of the country's terrain. Also, commanding American General William Westmoreland only knew how to wage a conventional war and not a guerilla one, which was another disadvantage for the U.S. military.

Consequent to North Vietnamese Foreign Minister Trinh's statement on January 28th regarding the American cease-fire, and before peace talks could officially begin, on January 31st, 1967, the U.S. bombing campaign against the north resumed where it left off. Yet by this time, the Vietnam conflict was draining $20 billion a year from the U.S. economy (pg 388, #5). However, this figure only represented 3% of the gross national product (G.N.P.). This figure was smaller than the Korean War figure of 12% of the G.N.P. and World War Two's 48%. Back at home, the American public began to protest its country's involvement in the war as early as 1966. But Johnson remained unaffected and committed to the war, saying that he would not allow North Vietnam to use "devastating aggression" to militarily succeed (pg 394, #5). Johnson would go on to describe the opponents of the war in the House as "nervous nellies."

By 1967, Johnson was picking targets to bomb in Vietnam himself, with the advice of Defense Secretary Bob McNamara. However, by this point, McNamara, along with McGeorge Bundy and Richard Goodwin, believed that the war could not be won (pg 397, #5). In fact, McNamara had his doubts about the U.S. winning the war as early as November 1965 (pg 494, #1). By 1967, after becoming convinced that the U.S. air strikes in the north of Vietnam were not succeeding, McNamara stated that "no amount of bombing can end the war" (pg 512, #1). This caused Johnson to drop McNamara from the Cabinet by February 29th, 1968, forcing his resignation. Yet Johnson would also lose Cabinet members McGeorge Bundy, George Ball, and Bill Moyers, who had all resigned due to their differences with the president over the war. McNamara was temporarily replaced as Secretary of Defense by Clark Clifford. National Security Adviser McGeorge Bundy was replaced by Walt Rostow.

On the home front, Johnson did implement some successful domestic policies in his Great Society programs, including raising the minimum wage from $1.25 to $1.60 an hour and establishing the Department of Transportation on October 15th, 1966 (pg 401, #5). Johnson continued his environment-friendly agenda by signing the Clean Waters Restoration Act and establishing new national parks. Johnson pledged federal money to combat rising crime and enacted tougher gun laws (pg 411, #5). Yet during this time, the federal deficit ballooned to three times the $1.8 billion projection. The Vietnam War was beginning to crowd out the Great Society programs fiscally, and the federal deficit grew from balanced in 1965 to $3.8 billion in 1966, to $8.7 billion in 1967, to $28 billion in 1968 (pg 442, #5).

Back in Vietnam, the situation for U.S. troops was only getting worse. General Westmoreland's strategy to win the Vietnam conflict was not working. Accordingly, by August 1967, Johnson sent an additional 47,000 troops to Vietnam (pg 424, #5). In South Vietnam, on September 3rd, 1967, General Nguyen Van Thieu was elected president. However, this would not stop the Communists, who began major military actions against the south that same month. The remaining members in Johnson's Administration told him to reject Robert McNamara's peace proposals, and to relentlessly go after his Communist military opponents.

However, on September 29th, 1967, Johnson said in a speech that he delivered in San Antonio, Texas, that the United States would stop bombing North Vietnam in exchange for "productive discussions" with North Vietnamese leaders. By 1967, more bombs were dropped on North Vietnam than during the entire World War Two campaign. Also by the end of 1967, more than 9,000 U.S. combat troops had been killed, out of a troop population of more than 500,000. General Westmoreland stated in November 1967 that he was optimistic about winning the war, and was publicly boasting that he was waiting for a Communist offensive to begin because he was looking for a fight from the north. On January 31st, 1968, General Westmoreland got exactly that.

On that day, the Vietcong Army launched the surprise Tet Offensive, which violated a holiday cease-fire agreement by attacking every military installation in South Vietnam and an additional 100 cities. The Vietcong attack was led by Vo Nguyen Giap, and it came as a complete surprise (pg 1004, #8). The Tet holiday was a celebration of the lunar new year (pg 530, #1). This was a holiday for both the north and south of Vietnam. The north used this U.S. military tactical mistake to launch a major attack on the south. The Vietcong even battled their way to the U.S. Embassy building. A Vietcong suicide team of 19 people blew themselves up outside the embassy, which killed seven U.S. soldiers (pg 447, #5). This surprise attack by the north was described as an intelligence failure by the United States that ranked with Pearl Harbor (pg 556, #1).

Despite this, the Vietcong did sustain massive casualties, which caused President Johnson to proclaim victory. The Tet Offensive was viewed by the Vietcong as a missed opportunity for victory (pg 558, #1). But the Tet Offensive proved to America that the war was becoming very unpopular with American citizens, and this ultimately hurt Johnson's popularity among the people. Johnson's approval ratings dwindled from 48% to 36%. His approval rating of handling the war itself went from 46% to 26% (pg 59, #1). These facts caused Johnson to seriously recalculate his position in running for re-election as president, with the election of 1968 coming up. He also took into consideration that he would be running against the brother of his slain predecessor, Senator Robert Kennedy of New York, who had adopted an anti-war stance against Vietnam the year before.

During the New Hampshire primary, Senator Eugene McCarthy of Minnesota nearly defeated Johnson by coming in a close second. After the president considered that Johnson men do not live into their mid-60s, and with the unpopularity of his views in the Vietnam conflict, Johnson concluded that his heart was not into running for president again in 1968. As a result, on March 31st, 1968, President Johnson addressed the nation on television to announce a partial bombing halt on North Vietnam, offer peace talks between the two nations, and announce, to a stunned nation, that he would not run for re-election as president. After his presidential term was completed, Johnson admitted that his failed policies in Vietnam cost him a second term as president.

After this announcement, Johnson was universally praised for his decision. The president was upset by the number of kids who were protesting against him and hated him for his policies. Johnson used to say, "Don't they realize that I am really one of them?" Johnson's withdrawal from the 1968 presidential race helped his stature in Congress and helped get his legislation passed, which included the passage of Johnson's Fair Housing Act on April 10th, 1968. Also, when Johnson announced his decision not to run for president again, Hanoi immediately opened peace negotiations in Paris (pg 349, #6). However, these talks that sought quick resolution of the war ultimately did not succeed in meeting their objectives.

Meanwhile, back in America, there were problems mounting because of tragedies that were occurring due to civil unrest. On April 4th, 1968, Dr. Martin Luther King, Jr. was assassinated in Memphis, Tennessee, which caused violent riots to take place in multiple American cities. Also, Democratic front-running candidate Senator Robert Kennedy was assassinated in Los Angeles, California, on June 5th, 1968, after he had just won the California primary. Vice President Hubert Humphrey would go on to claim the Democratic nomination for president in Chicago, Illinois at the convention, despite intense rioting and violent clashes with police outside the convention hall.

On the Republican side, former Vice President Richard Nixon won the nomination for the second time as the presidential candidate in Miami, Florida, on August 8th. Meanwhile, the Democrats were in disarray, with the assassination of Robert Kennedy in Los Angeles, the convention riots in Chicago, and Democratic candidate Humphrey receiving the brunt of criticism over Vietnam. Humphrey felt the political pressure enough that in September 1968, he announced that if elected, he would stop the bombing of North Vietnam as "an acceptable risk for peace" (pg 596, #1). To further Humphrey's chances of being elected president, Johnson announced on October 31st, 1968, that he was suspending all bombing of North Vietnam.

Despite this, it was not enough to propel Hubert Humphrey to the White House; on November 5th, 1968, Richard Nixon and Spiro Agnew were elected president and vice president of the United States. When Johnson finished his term on January 20th, 1969, he had served 32 years in politics as a Congressman, Senator, vice president, and president (pg 352, #5). Johnson would go on to build his library in Texas after his presidency, and he lived on his ranch. Johnson even grew his hair long. By the end of 1968, the American troop strength in Vietnam was 540,000.

On January 20th, 1969, Richard Nixon was sworn in as the 37th President of the United States. He would then inherit the Vietnam conflict and direct its outcome. Nixon was born on January 9th, 1913, in Yorba

Linda, California, and he graduated first in his class at Whittier High School. Nixon would go on to attend Whittier College and Duke Law School. Nixon graduated from Duke in 1937 (pg 583, #9). Nixon served in the U.S. Navy during World War Two, where he was stationed in the Pacific theatre. After his discharge in 1946, Nixon served in Congress from 1947 to 1950, the U.S. Senate from 1951 to 1953, and as the vice president to President Eisenhower from 1953 to 1961 (pg 586, #9).

President Nixon's Cabinet consisted of Vice President Spiro Agnew, Secretary of State (and former Attorney General for President Eisenhower) William Rogers, Defense Secretary Melvin Laird, Attorney General John Mitchell, Secretary of Labor George Schultz, and National Security Adviser Henry Kissinger, who would personally go on to play a very influential role in both the administration and the Vietnam War. The Soviet premier in 1969 was Alexei Kosygin (pg 144, #11).

Kissinger was born on May 27th, 1923, in Bavaria, Germany (pg 20, #7). His real first name is Heinz, but he changed his name to Henry when he immigrated to the United States. His family's religious faith is Judaism, and during his youth, many Jews were becoming ostracized in Germany, and their citizenships were revoked. These atrocities committed by the Nazis against Jewish citizens were called the Nuremberg Laws. While the Kissinger family successfully fled Germany to the United Kingdom on August 20th, 1938, and then to the United States, many members of his extended family were killed in the Nazi concentration camps (pg 29, #7). Upon Henry's arrival in the United States, he attended City College in New York and achieved excellent grades.

Kissinger served in the U.S. Army during World War Two and was selected to be in the Army Specialized Training Program. He served brilliantly for America in the European theatre and received a Bronze Star for his bravery. Upon his return to the United States after the war, Kissinger attended Harvard. During Nixon's first term in office, Kissinger had more influence on foreign affairs than Secretary of State Rogers (pg 98, #4).

By this time, the American commander in Vietnam was General Creighton Abrams, who replaced General Westmoreland. Abrams worked with the then-South Vietnamese President Nguyen Van Thieu (pgs 87-88, #4). Nixon described Thieu as "the most impressive South Vietnamese leader I've met." However, Nixon would also note, "that is not saying a great deal." Kissinger concurred with this opinion during his own visit to Vietnam in October 1965, in which he found both Thieu and the South Vietnamese government to be inept and corrupt (pg 118, #7).

Not helping matters for the American side in this war was the discovery of evidence related to the My Lai Massacre on March 16, 1968, where U.S. troops, under the command of Lieutenant William Calley, massacred innocent civilians, including women and children. Calley was court-martialed for his role in this awful incident and was later jailed and released.

Upon his acceptance of the presidency, Nixon wanted to end the Vietnam War with an armistice that was similar to that previously used to end the Korean conflict almost 20 years earlier. Nixon promised Thieu that during his first term as president, he would provide military support to the South Vietnamese Army, and then economic assistance during his second term (pg 89, #4). In Vietnam , 31,000 American troops had already died, and there were still 200 soldiers dying there every week (pg 235, #11). Nixon also had to deal with China, which was home to one fifth of the world's population, and China's rifts with Russia, which the United States was looking to exploit.

Nixon began secret U.S. military bombings of Vietnam's neighboring country, Cambodia, on March 18th, 1969. That same month, Secretary of Defense Melvin Laird invented the term "Vietnamization" as a rationale for American troop withdrawals. By June 1969, the U.S. military presence in Vietnam was over 500,000, and the casualty list at the time was an astronomical 35,000 American soldiers killed and over 200,000 injured. Feeling this pressure, Nixon proposed the withdrawals of both American and North Vietnamese forces from South Vietnam on May 14th, 1969. The next month, on June 8th, 1969, Nixon and Thieu announced, on the island of Midway, the withdrawal of 25,000 U.S. troops from Vietnam. Nixon specifically selected diplomatic heavyweight Kissinger to be National Security Adviser and deal with such foreign powers as China, the Soviet Union, and North Vietnam.

Nixon notoriously distrusted the State Department, the Pentagon, and the C.I.A., and he wanted all foreign policy decisions to come directly from the White House (pg 602 #1). Accordingly, Kissinger met covertly with North Vietnamese negotiator Xuan Thuy in Paris. The U.S. taxpayers were also paying $30 billion a year for this war. Kissinger analyzed that North Vietnam would win the war because it could sustain far more casualties than the Americans could, and would win so long as it did not lose (pg 603, #1). Despite this, Kissinger advised Nixon against a total withdrawal, because U.S. foreign policy had to maintain credibility (pg 160, #7). Kissinger began a four-year policy of a negotiated settlement with the North Vietnamese, and Nixon wanted Soviet help with these negotiations (pg 166, #7). However, the Soviets did not really put that much pressure on North Vietnam, despite the U.S. request for them to do so. By the spring of 1969, both sides of Congress were calling for troop withdrawal from Vietnam (pg 609, #1).

While Nixon had ruled out absolute victory in Vietnam by this point, he also refused to contemplate defeat. Nixon wanted to bomb North Vietnam's supply routes from Cambodia (pg 171, #11). It was Kissinger who found a pro-Communist camp over the Cambodian border (pg 172, #11). Accordingly, on March 18th, 1969, the U.S. military bombed

Cambodia. Nixon envisioned a durable peace rather than just an armistice (pg 604, #1). The code names for attacks over Cambodia were Lunch, Snack, Dinner, Dessert, and Supper (pg 176, #11). The mission was kept in secret, which angered the American public (pg 179, #11). The reason that President Nixon and Kissinger wanted to expand the war into Cambodia was that the North Vietnamese troops had organized there before the Tet Offensive, and there was evidence that they were building up infantry units in that location.

Kissinger's role was elevated in June 1969 to the number-two position on foreign policy, much to the annoyance of Secretary of State Rogers (pg 203, #7). In fact, Kissinger had his own connection to Soviet Ambassador Anatoly Dobryrin. As a result, Kissinger and Rogers often battled over ideas and policies. Both Nixon and Kissinger thought South Vietnamese President Nguyen Van Thieu's authority to be weak (pg 610, #1). But Kissinger did believe that North Vietnam had its breaking point, and that sustained bombing of North Vietnamese posts would get them to respect Thieu's power (pg 611, #1).

On September 2nd, 1969, Ho Chi Minh died at the age of 79 (pg 612, #1). Pham Von Dong took control of North Vietnam in Minh's place (pg 1019, #8.). Yet in America, massive anti-war demonstrations occurred in Washington, D.C. on October 15th, 1969. Nixon delivered his "silent majority" speech on November 3rd, 1969, as he asked the American public to be patient as he negotiated an "honorable end" to the war by withdrawing troops and replacing them with South Vietnamese troops (pg 615, #1). Yet according to President Nixon, North Vietnam had to respect the Saigon government.

By December 1969, American troop strength was reduced by 60,000 men in Vietnam. However, 30,000 U.S. troops had already died in Vietnam when Nixon took office, and another 10,000 soldiers died during Nixon's first year (pg 616, #1). On February 20th, 1970, Henry Kissinger began secret talks in Paris with North Vietnamese negotiator Le Duc Tho. During this time in Cambodia, its leader, Norodom Sihanouk, was overthrown by Lon Nol and Sisowath Matak on March 8th, 1970. These two oppressors helped to lead the Cambodian Communist movement known as the Khmer Rouge. This movement was responsible for slaughtering thousands of innocent Cambodians and Vietnamese people for political purposes. Accordingly, Cambodia was invaded by the U.S. military in May 1970, and President Nixon made this decision without Congressional authority; 31,000 U.S. troops and 43,000 South Vietnamese troops then entered Cambodia. Nixon compared this idea to President Teddy Roosevelt's invasion of San Juan Hill, although Kissinger and Secretary of Defense Melvin Laird did not agree with Nixon's decision (pg 269, #11).

In April 1970, 20,000 U.S. troops attacked two Vietcong bases in Cambodia as Nixon spoke to the country on television (pg 624, #1). The

White House wanted to eliminate North Vietnamese bases in Cambodia (pg 256, #11). According to Kissinger, the U.S. assault and invasion of Cambodia was enormously successful in capturing stockpiles of weapons and disrupting supply lines. This severely weakened the North Vietnamese military in Cambodia, and American G.I. fatalities dropped as a result. The place that was assaulted by U.S. forces in Cambodia was called Parrot's Beak (pg 570, #11). But the North Vietnamese military forces stayed one step ahead of U.S. troops by abandoning camps weeks before they were bombed (pg 625, #1). To the American public, Nixon seemed to be expanding a war that he had pledged to end, and there were widespread protests in America as a result. On May 4th, 1970, nervous National Guardsmen opened fire on protesting students at Kent State University in Ohio, resulting in the deaths of four students. As a result, riots exploded nationwide, and demonstrations occurred on 1,100 campuses (pg 216, #4).

Despite the domestic protests, Nixon proposed a "stand-still cease fire" on October 7th, 1970, but he continued to repeat his mutual-withdrawal formula the next day. During this time, Nixon did succeed in getting American combat deaths to their lowest totals (24) since October 1965. The American troop numbers in Vietnam were down to 280,000 by the end of 1970. By February 1971, the South Vietnamese military began maneuvers in Laos against the Ho Chi Minh Trail. The aforementioned Lieutenant Calley was found responsible for the My Lai Massacre and was convicted of pre-meditated murder of South Vietnamese citizens on March 29th, 1971. Calley did serve jail time, but was later released.

At home, the *New York Times* began publishing the "Pentagon Papers" on June 13th, 1971, by Daniel Ellsberg, who was a Harvard scholar and former Marine captain in Vietnam (pg 332, #4). Ellsberg's ideology transformed into a very anti–Vietnam War platform upon his return home. Because of this press story, Nixon cut off all ties with the *New York Times* (pg 333, #4). Nixon desperately wanted to plug the leaks that came out due to Ellsberg's publication of the Pentagon Papers, and he was willing to do anything to achieve that result (pg 338, #4). As a result, Nixon's Cabinet adviser, John Ehrlichman, and Chief of Staff H.R. Haldeman organized the "plumbers" operation to investigate Daniel Ellsberg, who was responsible for making the Pentagon Papers public.

Nixon ordered the break-in of the Brookings Institution for information on Ellsberg (pg 340, #4). Nixon himself was aware of the plumbers operation as early as 1971 (pg 369, #4). The first break-in was called Hunt-Liddy #1, which occurred at the Los Angeles office of Dr. Lewis Fielding, who was Ellsberg's psychiatrist . This break-in took place on August 3rd, 1971. The men involved in the operation then were paid $5000 in cash from President Nixon's campaign funds.

Back in Vietnam, Nguyen Van Thieu was re-elected President of South Vietnam on October 3rd, 1971. By December 1971, American troop strength in Vietnam was down to 140,000. By January 25th, 1972, Nixon revealed that Henry Kissinger had been secretly negotiating with the North Vietnamese. Nixon also announced that 45,000 more troops would be withdrawn from Vietnam by February 1st, 1972. This withdrawal would leave the same number of U.S. forces in Vietnam as at the end of 1965 (pg 388, #4). In the beginning of 1972, Nixon said that the final U.S. withdrawal from Vietnam should coincide with the complete release of all U.S. P.O.W.s (pg 418, #4). By early 1972, the U.S. military had successfully withdrawn over 410,000 troops from Vietnam (pg 1046, #11).

During this time, Nixon was making overtures towards China to strengthen ties between the two nations, and to secretly meet with Chinese Premier Mao Tse-tung. Nixon believed that China's willingness to talk to the United States was mainly due to China's concern about the Soviet Union and Japan (pg 415, #4). Nixon would learn that the Soviets felt the same way towards the Chinese. On March 30th, 1972, North Vietnam launched an offensive across the demilitarized zone. In response, on April 15th, 1972, Nixon authorized the bombing of an area near Hanoi and Haiphong.

Meanwhile, Attorney General John Mitchell resigned his position in 1972 in order to take over President Nixon's re-election campaign. Assistant Attorney General Richard Kleindienst took over for Mitchell (pg 432, #4). In May 1972, Nixon made great strides in negotiating limits to the arms race, and he and Soviet leader Leonid Brezhnev signed two agreements at the Strategic Arms Limitation Talks, called S.A.L.T. In this agreement, the United States and the Soviet Union were limited to two anti-ballistic missile sites, one to protect their countries' capital cities and the other to guard an offensive missile site, which was located approximately 800 miles away (pg 596, #9). The second portion of the agreement was for both nations to stop accumulating the numbers of strategic offensive ballistic missiles and keep them at approximately their current levels as of May 1972. This was a monumental breakthrough in improving U.S.-Soviet relations, and a major diplomatic victory for both Nixon and Kissinger.

However, back in Vietnam, the Vietcong captured the city of Quangtri on May 1st, 1972. Nixon responded by announcing the U.S. mining of Haiphong Harbor and intensification of the bombing of North Vietnam. The North Vietnamese Premier in 1972 was Pham Van Dong (pg 1109, #11). Yet at this time, a seemingly meaningless break-in at the Democratic National Committee headquarters at the Watergate complex in Washington would sow the seeds of the resignation of President Nixon and the imprisonment of most of his Cabinet. On June 17th, 1972, five men who worked for the Committee to Re-Elect the President, Bernard Baker, James McCord, Eugenio Martinez, Frank Sturgis, and Virgilio Gonzalez, were arrested for breaking into the Democratic National Committee's offices at the Watergate complex in Washington. The cover-ups by the president and his Cabinet that followed would ultimately lead to President Nixon's resignation in August 1974. Meanwhile, Kissinger met again with Le Duc Tho, the North Vietnamese negotiator, on August 1st, 1972. Kissinger sensed that the peace process had improved as a result of these meetings. But South Vietnamese leader Thieu was reluctant to accept the cease-fire accord and opposed the peace draft agreement formulated by Kissinger and Tho. This was because the agreement between Kissinger and Tho would allow North Vietnamese troops to remain in the south, but Thieu would remain in power (pg 453, #7).

Accordingly, the last U.S. ground combat troops were withdrawn in August 1972, and only military support personnel remained in Vietnam to guide the South Vietnamese troops. The Paris meetings succeeded in achieving significant breakthroughs between Tho and Kissinger on October 8th, 1972. Yet South Vietnamese leader Thieu still adamantly opposed the agreement. Thieu did not want U.S. military forces to leave Vietnam, so he became uncooperative in the final stages of the negotiations between Kissinger and Le Duc Tho (pg 1411, #11). One of Kissinger's many strengths was as a bargainer, and this was especially true in the Vietnam settlement. Hanoi Radio began to broadcast details of the peace agreement on its station in an effort to place pressure on both Kissinger and the United States. When North Vietnam was in the final stages of settlement in October 1972, it dropped the demand for a coalition government in South Vietnam, saying "the political struggle of South Vietnam [is] left for the Vietnamese to settle" (pg 1345, #11). However, Kissinger remained anxious to communicate to North Vietnam's leaders that, in fact, "peace is at hand," despite publicly making this statement to the press. Contrary to this notion, Nixon knew that it was not. Kissinger desired that the Vietnam War end before the November 1972 presidential election, with President Nixon running against Senator George McGovern. Technically, this did not happen, but President Nixon was re-elected in a landslide over Senator McGovern anyway on November 7th, 1972. Kissinger resumed talks with Le Duc Tho on November 20th and presented him with the 69 amendments to the agreements demanded by Saigon and Thieu. They would not get them.

Fresh talks between Kissinger and Le Duc Tho began again in December 1972, but broke down. As a result, on December 18th, 1972, Nixon ordered the "Christmas bombing" of North Vietnam, which involved 129 B-52 bombers, F-111s, and A-6 fighter bombers (pg 553, #4). This act killed 1,318 people in Hanoi and an additional 300 in Haiphong (pg 469, #7). While only military targets were supposed to be hit in this attack, the Indian and Egyptian embassies were also struck by the bombings. Though this military act was indeed controversial, it did bring North Vietnam back to the bargaining table (pg 555, #4). Even Pope Paul VI denounced the bombings. Kissinger, as chief negotiator, was against them also (pg 556, #4). Despite this, Nixon felt that the Christmas bombing was necessary because the Democratic Congress that was about to be sworn in threatened to cut off all funding for military operations in Vietnam. As a consequence, Nixon felt that he needed an agreed settlement from North Vietnam, which would be crucial in finally ending the conflict.

However, this bombing was intended to force the North Vietnamese to settle the war more favorably for the United States (pg 667, #1). A treaty was arranged between the United States and Vietnam that was formally signed on January 27th, 1973, officially ending American participation in the Vietnam War. When President Nixon announced this cease-fire, he said that within 60 days of January 27th, 1973, all 23,700 troops and P.O.W.s would come home to America. South Vietnamese leader Thieu was forced to accept the conditions of the U.S. agreement, because if he did not, the United States would cut off all aid to South Vietnam (pg 562, #4).

In fourteen years of American involvement in Vietnam, 8.7 million American soldiers went to Vietnam, and 58,151 soldiers were killed in this conflict (pg 564, #4). Three million Vietnamese people died in the Vietnam War (pg 354, #2). While the American involvement in the Vietnam War was over, by 1974, Thieu had declared that the war had begun again, but Nixon was too caught up in the Watergate scandal to concern himself about sending troops back to Vietnam. By the spring of 1974, North Vietnam had recaptured all the territory it had lost in the Mekong Delta to the south following the truce (pg 675, #1). By April 30th, 1975, Saigon had fallen to the North Vietnamese forces, and they had successfully captured the whole country. Kissinger blamed the Watergate scandal and the subsequent resignation of Richard Nixon as contributing to South Vietnam's collapse in 1975 (#11). History shows us that communism was contained in Berlin and Korea, but not in Vietnam (pg 1079, #8). However, Bob McNamara stated in his book, *In Retrospect*, that the Vietnam War did succeed partially in containing the spread of communism. McNamara went on to say that the lessons of Vietnam can be summed up by concluding that wars should be avoided, except when our national security is clearly and directly threatened (pg 332, #2).

Bibliography

1. *Vietnam: A History*, by Stanley Karnow (winner of the Pulitzer Prize) (Viking/Penguin, 1984), disc 5.
2. *In Retrospect: The Tragedies and Lessons From Vietnam*, by Robert McNamara, with Brian Van de Mark (Random House, 1996), disc 10.
3. *Kennedy: Profile in Power*, by Richard Reeves (Simon & Schuster, 1993), disc 5.
4. *President Nixon: Alone in the White House*, by Richard Reeves (Simon & Schuster, 2002), disc 5.

5. *LBJ: A Life*, by Irwin and Debi Unger (John Wiley and Sons, 1999), disc 4.

6. *Lyndon Johnson and the American Dream*, by Doris Kearns Goodwin (St. Martin Griffin, 1991), disc 3.

7. *Kissinger*, by Walter Isaacson (Simon & Schuster, 1992), disc 3.

8. *The Wars of America*, by Robert Leckie (previously listed).

9. *The Complete Book of U.S. Presidents*, by William A. Degregario (previously listed).

10. *The Killing Zone: My Life in the Vietnam War*, by Frederick Downs (W.W. Norton, 1978).

11. *White House Years*, by Henry Kissinger (Little, Brown, 1979).

12. *Kennedy*, by Theodore C. Sorenson (William S. Konecky Associates, 1999).

INTERVIEW WITH DR. ROBERT KEIGHTON

Thomas Athridge: Did LBJ really have an understanding of the lives of poor people when he was teaching poor children at Wellhausen and Sam Houston High School, and which he later used in his years as president?
Robert Keighton: Yes, this did have an effect on him. He had genuine concern for the poor, the blacks, and Mexicans, and it influenced him. He was like a political chameleon.
TA: How much influence did LBJ have during his years in Congress, before his Senate tenure?
RK: Although LBJ was a protégé of Sam Rayburn, I don't think that he had a great amount of influence. He was more of a power broker in the Senate than in the House.
TA: Did the lessons LBJ learned during the World War Two conflict ultimately cause his downfall during Vietnam?
RK: Yes, it was a factor, but not the only one. LBJ was advised by bright people in his Cabinet who wanted to escalate the conflict. LBJ was also getting inaccurate information on the casualty rate in Vietnam. He was stuck in a catch-22 situation.
TA: How unique was it that Johnson went off to war as a member of Congress?
RK: Reasonably unique, but others have probably done it, also.
TA: How popular or unpopular were LBJ's views on civil rights in the South at the time?
RK: When he was president, they were unpopular views, certainly, at the time. In fact, Texas had segregated schools when LBJ became vice president.
TA: How was LBJ's relationship with Eisenhower during LBJ's reign as Senate majority leader?
RK: Johnson was enormously powerful as majority leader, and Republicans had to work with him.

Chapter 13
The 1991 Gulf War: What Led the United States to War, and the Actions and Decisions of President George H.W. Bush

The Gulf War that began on January 15th, 1991 and lasted for six weeks, between U.S.-led forces and the Iraq military, resulted in a decisive U.N. coalition victory that ousted Saddam Hussein's Iraqi military from occupying Kuwait. This was the first war that the United States had been in since the Vietnam War ended in 1973. In that space of time, however, events on the global stage had the United States fighting terror internationally.

In 1978, Ayatollah Khomeini, an Islamic extremist, who was then in exile in Paris, returned to power in Iran after the overthrow of the Shah, and consequently held 52 Americans from the U.S. Embassy hostage for 444 days. The hostages were released on January 20th, 1981, which was the day Ronald Reagan was sworn in as our 40th president. Later, on October 23rd, 1983, The U.S. Marine barracks in Beirut, Lebanon was attacked by a suicide bomber who was armed with 12,000 pounds of dynamite, and he murdered 241 Marines who were mostly asleep at the time. This act of terror caused the withdrawal of the Marines from Lebanon.

On September 1st, 1983, a Soviet missile mistakenly shot down a Korean airliner, because the Soviets thought that it was an enemy plane. This act killed 269 people (pg 284, #1). Unfortunately, the United States would make the same mistake in 1988, when the *U.S.S. Vincennes* mistakenly shot down an Iranian flight, which killed 290 innocent people (pg 285, #1).

In 1985, President Reagan launched an attack on Grenada in order to remove the Soviet-Cuban influence. Reagan did so successfully. However, Reagan would later get into some trouble when members of his Administration sold weapons to Iran in an effort to normalize relations between the two nations, and used the profits from these sales to fund the Nicaraguan Contra rebels to fight the Communists.

Meanwhile, Iran and Iraq were involved in a brutal war, begun by Saddam Hussein's order for the Iraqi military to invade Iran in 1980. This war, which would prove to be a stalemate, was extremely costly to both countries in lives and money (especially to the Iraqi economy). In Hussein's case, he was once supported by U.S. funding and the use of military equipment to fight the Iranians, but by the end of the eight-year conflict, he was in debt by $90 billion (pg 14, #3). Also during this time, Hussein had used chemical weapons against both the Iranians and his own people, the Kurds. These chemical weapons were unintentionally provided by the Americans.

To solve Hussein's economic crisis, he decided to invade Kuwait in order to seize and control its oil fields. Accordingly, in an act of total hostility, Hussein's Iraqi forces invaded the country of Kuwait on August 2nd, 1990, and seized control of the whole country in one day. This act of aggression by Iraq led President George H.W. Bush and his allies, including British Prime Minister Margaret Thatcher and Soviet Premier Mikhail Gorbachev, to decide that Hussein and his troops had gone too far and must pull out of Kuwait. This was stated in UN Resolution 660. UN Resolution 660 also stated that if Iraq did not pull out of Kuwait, then the consequence would be war.

Clearly, Hussein had his sights on Saudi Arabia's oil fields, and unquestionably would have gone after them next had the Allied forces not intervened. This war was a result of an act of aggression that Saddam Hussein, a former ally of the United States, perpetrated by launching a premeditated and unprovoked attack against its Kuwaiti neighbor, to which the United States felt obligated to react. This decision for military intervention was made by President George H.W. Bush. Yet in order to understand why the first President Bush led our nation to war against an enemy far away in a desert climate, we must first analyze Bush the leader, and what led him to embark on his fateful journey to becoming a war president.

President George Herbert Walker Bush was born on June 12th, 1924 in Milton, Massachusetts. Bush's parents were Dorothy and Senator Prescott Bush, who represented Connecticut. George got his name from his maternal grandfather, who was named George Herbert Walker. In fact, his grandfather Walker obtained a massive fortune from his banking investment firm, Ely and Walker Company, which was one of the biggest in the country (pg 23, #4).

Before George attended Yale University upon his graduation from high school, he decided to enlist in the U.S. Navy, due to World War Two conflict, in which the United States was involved at the time. While in the Navy, George attended flight school, and he became the youngest fully fledged commissioned pilot in the Naval service (pg 47, #4). Bush became a squadron leader and bombed Japanese targets off the coast of Saipan and Tinian. He logged 1,228 hours of flying time as a pilot, completed 126 carrier landings, and flew 58 missions, on two of which he was shot down but successfully landed (pg 63, #4). Bush garnered the distinguished Flying Cross for his combat bravery (pg 58, #4).

Upon his return home from the war, George and his wife Barbara had their first child, future President George W. Bush, on July 6th, 1946. Then, Bush attended Yale University, where he majored in business and economics (pg 65, #4). After he successfully graduated from Yale, George and Barbara had an additional five children: John Ellis, or Jeb, Neil Mallon, Marvin Pierce, Dorothy Pierce, and Robin. Robin tragically died of leukemia in 1953.

Bush went into the oil business with a company called Zapata Oil, which was located in Midland, Texas. Bush was successful in finding oil deposits there (pg 82, #4). The city of Midland profited heavily as a result of this discovery. By 1959, Bush's personal stock of Zapata's offshore accounts comprised approximately 15% of the company, equivalent to $600,000. Bush began a life-long and profitable friendship with brothers Bill and Hugh Liedtke, who, together with Bush, created the company of Pennzoil. They did this by combining Pennzoil with Zapata Oil (pg 85, #4). Bush was a self-made millionaire in the oil business by the age of 35 (pg 87, #4).

Though Bush became a very successful businessman and war hero, he had an urge to follow politics, like his father, Prescott. His father was a Republican Senator from Connecticut. Prescott retired from politics in 1962 (pg 89, #4). So, on September 11th, 1964, George H.W. Bush decided to run for a U.S. Senate seat in Texas, against Democrat Ralph Yarborough, the incumbent. Bush's campaign focused on "cleansing the liberals from Washington" (pg 109, #4). Bush even impressed former Vice President and future President Richard Nixon with his vigorous campaign style, but Nixon still believed that Bush would lose to the incumbent. Accordingly, Senator Yarborough defeated Bush by 300,000 votes in the November 1964 election, which also saw President Lyndon Johnson win his race in a landslide. Though Bush and President Johnson represented different parties and points of view, the two men respected each other due to their Texas ties. This might explain why Bush was such a staunch supporter of the Vietnam War. Bush believed that the war against the Communists could be won with patience (pg 130, #4).

Though Bush lost his bid for a Senate seat in 1964, he did win a Congressional seat in 1966 (pg 121, #4). While in Congress, he was appointed to the powerful House Ways and Means Committee (pg 125, #4). Bush opposed the Civil Rights Act of 1964, but later supported the Civil Rights legislation enacted in 1968. Bush ran unopposed for his Congressional seat in 1968 (pg 133, #4). During this time, Bush surprisingly favored abortion rights for women, and he explained that the right to an abortion "should always remain a matter of individual choice" (pg 134, #4).

In 1970, Bush ran for Yarborough's Senate seat again; this time, it was against Lloyd Bentsen, who beat the incumbent Yarborough in the primaries. In the battle of Bush versus Bentsen for the Senate, Bentsen defeated Bush, who lost his bid for the Senate a second time. President Nixon, however, stepped in to offer Bush the position of Ambassador to the United Nations, which he accepted (pg 148, #4). Nixon wanted Bush to be his point man within the extended international community in New York. It was Bush who got to criticize India on behalf of the Nixon Administration concerning the India/Pakistan situation. Both Nixon and Kissinger favored Pakistan, in part because they hated Indira Gandhi (pg 153, #4).

By 1972, Nixon was competing for his re-election as president against Senator George McGovern. Nixon would eventually win this election in a huge landslide. At the Summer Olympics in Munich, however, Palestinian terrorists massacred all of the Israeli athletes during Bush's tenure as Ambassador to the United Nations. During this time, Nixon and Kissinger were attempting to orchestrate a cease-fire with North Vietnam, which eventually transpired in January 1973. Bush's father Prescott died in 1972, also.

During 1973, Nixon's Chief of Staff H.R. Haldeman, and advisers John Ehrlichman and John Dean, had to resign as a result of the Watergate scandal, so Nixon placed all damage-control assessment of his presidency on Bush (pg 161, #4). With Nixon's eventual resignation and Gerald Ford's rise to the office, President Ford named Bush head of the C.I.A. (pg 188, #4). While he was head of the C.I.A., he was initially an ally of Manuel Noriega of Panama, because Noriega would act as the intermediator for the United States and Fidel Castro's Cuba (pg 203, #4). In fact, Bush met with Noriega at the Panamanian Embassy to discuss issues on December 8th, 1976 (pg 204, #4).

During this time, Bush's ideology revolved around business, free markets, and free trade, and he would become the political spokesman for lower taxes, fewer regulations, gun freedom, and capital punishment (pg 212, #4). It was in 1979 that George H.W. Bush announced his candidacy for the Republican nomination for president (pg 217, #4). His main opponent was California Governor Ronald Reagan. Though Bush was victorious in the Iowa caucus, he quickly lost momentum to Reagan, and lagged far behind him in the polls.

Despite this, Bush showed the Republican Party he could be a solid representative of their issues, and had support from a number of constituents. Bush had help from Lee Atwater, who influenced him and helped him stay connected to Reagan's team. At the 1980 Republican convention, the nominee Reagan had a choice of 3 vice-presidential candidates: James Baker, former President Gerald Ford, or George H.W. Bush. After Reagan almost went with Ford, controversy over Ford's media comments about his role in a Reagan administration swayed Reagan to

211

ask Bush to be his vice presidential nominee, to which Bush said yes. This pick by Reagan added strength to the ticket by uniting the Republican primary voting constituents.

On election night, in November 1980, the Reagan/Bush ticket defeated the Carter/Mondale ticket by winning 44 states and 489 electors to Carter's 6 states and 49 electors (pg 254, #4). This election was also successful in getting the Iranian hostages freed on Inauguration Day in 1981, after 444 days of captivity. Right after being inaugurated, Reagan put Vice President Bush on an international anti-terrorism committee (pg 265, #4).

The new administration was tested right away when, on March 30th, 1981, President Reagan was shot and almost assassinated by John Hinckley, Jr. Though Hinckley failed to kill the president, he did shoot and wound him, as well as Jim Brady and two others. During this chaotic time, Bush acted as he should have and did not necessarily take control, though he could have if he chose to do so (pg 271, #4). However, President Reagan fully recovered from his wounds.

After the previously mentioned Marine barracks bombing in Beirut, in which 241 Marines were murdered by a suicide bomber, Reagan chose to pull all U.S. forces from the region. To quell the international suspicion that the United States could be intimidated out of a geographic area, President Reagan chose to flex the U.S. military's might on a Soviet-Cuban colony called Grenada. According to Richard Clarke's *Against All Enemies*, Reagan invaded Grenada after the Beirut massacre to show that the United States could still win wars. This successful military campaign in Grenada restored the U.S. military's morale by succeeding in a short invasion of the country and restoring democracy to Grenada. In 1984, Reagan and Bush won re-election against the Democratic ticket of Mondale/Ferraro in a landslide, despite the nation's economy suffering from fiscal mismanagement (pg 294, #4).

In Reagan's second term (1985-89), his Administration was hurt by the Iran-Contra scandal that developed when Administration officials sold weapons to Iran and used the proceeds to fund the Contras in Nicaragua (pg 316, #4). This controversy hurt Reagan, Bush, and the Administration in the eyes of the people. Caspar Weinberger, then Secretary of Defense, was heavily criticized for his role, and he was indicted by an independent counsel. Weinberger would later resign as Secretary of Defense (pg 348, #1). Colonel Oliver North was also indicted by the grand jury but was later acquitted. Bush later pardoned Weinberger when he became president. Weinberger was replaced as Secretary of Defense by Frank Carlucci.

Vice President George H.W. Bush began his quest for the presidency by announcing his bid for the Republican nomination on October 13th, 1987, in Houston, Texas (pg 321, #4). By January 1988, Bush was 12 points behind in the polls to Senator Bob Dole (pg 323, #4). While questioning Bush on his involvement in the Iran-Contra scandal, CBS news anchor

Dan Rather and Bush got into a verbal confrontation, in which Bush questioned Rather's walk-out from a previous newscast for CBS, on a time delay from the U.S. Open tennis tournament. This seemed to show that Bush had a backbone to stand up against other people.

By the time of the New Hampshire primary, that state's Governor, John Sununu, helped Bush win the state. This momentum carried all the way to Super Tuesday, in which Bush won all of the remaining primaries. By April 16th, 1988, Bush had the Republican presidential nomination locked up (pg 330, #4). However, Bush's problems in winning the presidency from Democratic candidate Governor Michael Dukakis were far from over.

By May 1988, with both campaigns well under way, the polls showed that the Democratic candidate Dukakis was 16 points ahead of Vice President Bush. While the campaign progressed, the case of convicted killer Willie Horton was made public, which involved Dukakis's role as Governor of Massachusetts. In this case, convicted felon Willie Horton was given, by mistake, a weekend furlough from jail and subsequently kidnapped and raped a female hostage. It was pointed out by the Bush campaign that only the state of Massachusetts made this offer to felons who were serving life terms, and this hurt Dukakis in the polls and among voters.

Yet the post-convention bump for Democratic candidate Dukakis, which was as high as 17 points, had closed to 7 points by August. When it was time for Bush to pick a vice-presidential candidate, he had as his two finalists Bob Dole and Dan Quayle, who were both Republican Senators. After careful consideration by Bush, he chose Senator Quayle. Once Bush's pick was announced, Quayle's credibility was attacked, and he was accused of possibly being a draft dodger in the Vietnam War. Also at the convention, Bush's acceptance speech that contained the phrase, "Read my lips: no new taxes" would come back to haunt him in his eventual presidency. Yet Bush managed to climb back from a substantial poll deficit to Dukakis to overtake the Massachusetts Governor by the debates.

On election night, Vice President Bush won 53.4% of the popular vote and 40 states to win the presidency over Governor Dukakis. While celebrating his victory, Bush was faced with a Democratic House and Senate. Bush named James Baker as his Secretary of State (pg 356, #4). John Sununu was rewarded for his heavy campaigning by being named Chief of Staff. Marlin Fitzwater was the Press Secretary, Brent Scowcroft was named National Security Advisor (pg 360, #4). Condoleezza Rice was then on Scowcroft's National Security staff. She would later be the National Security Advisor to the second President Bush in his first term, as well as Secretary of State in his second. Colin Powell was named Chairman of the Joint Chiefs. Finally, Dick Cheney was named Secretary of Defense, but only after Bush's first pick, John Tower, was rejected by the Senate.

When Bush's Cabinet was finally in place, his first priority was to cut the deficit from the Reagan Administration. Second, Bush wanted to remove Soviet influence from Eastern Europe. Bush felt that a free and united Germany would greatly help the push for a more democratic and peaceful Europe (pg 390, #4). Meanwhile, Bush's former ally, Manuel Noriega, was charged with infringing on national elections in his country when Bush took office. Noriega's actions, according to Bush, were so outrageous to the peace process in Nicaragua that Bush decided to send troops to secure Noriega's ouster. This operation was set to begin on December 20th, 1989, and was named by the military as "Operation Just Cause" (pg 426, #1).

Right on time, December 20th saw the US military invade Panama. The Army Rangers landed at Rio Hato after they stormed Modelo Prison and rescued prisoners there (pg 428, #1). Meanwhile, the 82nd Airborne Division took control of Torrijos Airport, while the 7th and 82nd Airborne Divisions took the city of Colon after some heavy fighting. Four U.S. Marines died at Punta Paitilla Airport. The U.S. mission there succeeded in defeating Noriega's forces, and Noriega himself was forced to hide at the Papal Nunciatura in Panama City. After the United States had him surrounded and blared loud rock music to get him out, Noriega finally surrendered on January 3rd, 1990. He was charged as an international war criminal.

However, the United States was censured by the United Nations and the Organization of American States for invading Panama. The invasion was popular, however, with the Panamanians, when democracy was restored there. Also, the American people supported the invasion. The House Armed Services Committee investigated the invasion and found that 100 Panamanian civilians and 200 soldiers had died in the conflict. President Guillermo Endara then began his term. General Powell supported Bush's decision to invade Panama by stating that decisive force ends wars quickly and, in the end, saves lives (pg 434, #1).

In total, of the 24,000 U.S. troops who participated in the invasion of Panama, twenty-three were killed and 322 were wounded. Meanwhile, the Berlin Wall fell at the end of 1989, and the two Germanys were reunited. Bush and Gorbachev worked to reduce nuclear weapons by signing the 1991 START (Strategic Arms Reduction Treaty), which promised deep cuts in each nation's nuclear arsenal. By August 1990, a new threat appeared on the radar screen in the Persian Gulf, and that threat was Saddam Hussein's Iraq.

As stated earlier, Iraq invaded Kuwait on August 2nd, 1990 to possess its lucrative oil fields, because Hussein was desperate for money. Hussein also wanted access to the sea. Hussein badly miscalculated the world's reaction to his invasion. In fact, in Reagan's Administration and early on in George H.W. Bush's Administration, the United States and Iraq had

214

diplomatic ties to each other, including weapons sales to Iraq by the U.S. government (pg 694, #6).

Hussein also hurt his cause by personally lying to then-Egyptian President Hosni Mubarak, telling him that he would not invade Kuwait right before he did exactly that. It seemed clear that neighboring Saudi Arabia, with its vast and lucrative oil fields, would be Hussein's next target. It was also clear that something drastic needed to be done, and promptly.

Due to fortunate circumstances, the U.S. military was built up to a high extent due to preparations for the Cold War. The U.S. military had many forms of defense at its disposal. The first Bush Administration immediately gave the international community and the press a variety of reasons for going to war, including Hussein's capability of acquiring nuclear weapons and that, like Hitler, Hussein was a bloodthirsty dictator whose acts of aggression must be stopped sooner rather than later.

The top two military commanders, General Colin Powell, then Chairman of the Joint Chiefs of Staff, and General Norman Schwarzkopf, then head of US Central Command, began the battle plans for the military operation that would become Operation Desert Shield/Storm. The invasion would come from Saudi Arabia, which had given the United States permission to launch strikes from its soil in order to protect it from invasion.

On August 8th, 1990, Iraq declared the oil-rich country of Kuwait as a province of Iraq. Bush quickly marshaled support from the rest of the world to isolate Iraq from the world stage, and the United Nations subsequently condemned the invasion and ordered the Iraqi Army's immediate withdrawal. A trade and financial embargo against Iraq was also implemented. The allied coalition consisted of Argentina, Australia, Bahrain, Bangladesh, the United Kingdom, Canada, Czechoslovakia, Denmark, Egypt, France, Pakistan, Germany, Greece, Italy, Morocco, the Netherlands, New Zealand, Saudi Arabia, Spain, Syria, the United Arab Emirates, Pakistan, Kuwait, Oman, Qatar, Poland, and the United States. The Soviets, Greece, Norway, Spain, the Netherlands, Denmark, and Belgium sent naval assistance.

With everything militarily set in place from Saudi Arabia, U.N. forces issued an ultimatum to Hussein to withdraw from Kuwait by January 15th, 1991, or there would be war. Originally, Hussein ordered the round-up of approximately 3 million foreign civilians, including thousands of Americans, and placed them at strategic sites in Iraq and Kuwait in order to use them as human shields. Bush told Hussein that this tactic would not deter him from launching a war on Iraq, so Hussein blinked and released the prisoners (pg 695, #6).

Nevertheless, Operation Desert Storm was ready to be launched on Iraq, and when the deadline imposed on Hussein by the Untied Nations passed on January 15th, 1991, Bush gave the go-ahead for war. On January

17th, 1991, the coalition bombing campaign began. Those forces were commanded by General Norman Schwarzkopf, who was placed in Supreme Command by Bush. Bush was conscious of not repeating the mistakes of President Lyndon Johnson in the Vietnam War, who ordered the air strikes personally (pg 695, #6).

During the first wave of air strikes, stealth fighter jets knocked out Iraqi radar defenses around Baghdad, giving the coalition forces control of the skies for the rest of the operation. Despite this, the Iraqi Army had the fourth largest army in the world at the time (pg 17, #3). The coalition air strikes had devastated the Iraqi armed forces on the ground for over a month before the coalition ground assault began. In fact, for over six weeks, more than 2,000 coalition war planes dropped 88,500 tons of bombs over Iraq and parts of Kuwait, which destroyed roads and bridges, nuclear and chemical facilities, and conventional military targets (pg 695, #6).

In retaliation to these U.N. air strikes, Iraq launched Soviet-built Scud missiles at the country of Israel and its cities (pg 695, #6). Israel wanted to respond to these attacks militarily, but was convinced not to do so by the United States, which promised to protect Israel from Iraqi attack. Despite this promise, some Scud missiles did successfully strike targets in Israel, causing injury and death to some Israeli citizens. Patriot anti-missiles from the United States did intercept some of the Scuds. Iraqi mines were a distinct threat to U.S. Naval ships, due to the small proximity of the geographic sea area that they were patrolling (pg 42, #3). Therefore, it was crucial to sweep the area of all mines.

The army found the T.L.A.M.s (Tomahawk Land Attack Missiles) to be useful, offering the capability to attack a target deep in enemy territory without risking the loss of a valued pilot (pg 44, #3). The ground offensive took place on schedule to Bush's timetable, and on February 23rd, 1991, the assault began at 4:00 a.m.: 94,000 U.S. Marines were committed to the Gulf campaign, and they invaded Kuwait. The coalition campaign swiftly beat the Iraqi Army, yet there were some accidental deaths, with coalition aircraft mistakenly killing nine British soldiers and wounding eleven. There were also friendly-fire casualties for the U.N. forces, one of the most unfortunate aspects of any war.

General Schwarzkopf launched the ground offensive, called the "Hail Mary" strategy, in which the Marines tricked the Iraqis into believing that the United States was staging an amphibious assault. However, the main U.S. military force swept west, completely around the Iraqi defensive fortifications, to strike Hussein's Republican Guard and cut off the demoralized units on the front line. Nevertheless, the capital of Kuwait City was retaken by U.N. coalition forces on February 27th, 1991. This was the first complete victory by the United States in war since World War Two (pg 72, #3).

In retaliatory response to his inevitable defeat, Hussein gave the order to his evacuating forces to set all of Kuwait's oil fields ablaze. Yet once Kuwait was officially liberated from Iraq, Bush decided not to go to Baghdad to capture Hussein, and officially declared the war over with the official liberation of Kuwait.

Of the 541,000 U.S. troops who were deployed in the Persian Gulf War, 148 were killed and 467 were wounded, but 15% of the casualties came from friendly fire. Another 141 soldiers were killed from other U.N. nations that participated in the liberation of Kuwait. With Hussein and his forces defeated and Kuwait liberated, Bush's public approval rating soared to 89%, the highest such rating ever recorded in a Gallup poll. Bush had successfully beaten the "Vietnam syndrome."

Yet this 89% approval rating slipped all the way to 22% by mid-1992 (pg 497, #4). Despite this, the borders of Iraq and Kuwait were restored. Iraq had lost 80% of its tanks, 90% of its artillery, and 50% of its armored personnel carriers. United Nations Security Council Resolution 687 was then passed, which covered the issues of Kuwait's borders, war reparations, and the creation of a special U.N. commission to ensure the removal from Iraq of all biological and chemical weapons. This included the banning of all ballistic missiles with a range of over 93 miles, to which Iraq had agreed (pg 84, #3).

Operation Desert Storm restored faith in the U.S. military from the general American public (pg 1141, #5). Yet Bush's failure to help the economy and lower unemployment numbers would hurt him in the 1992 presidential campaign against Arkansas Governor Bill Clinton. Clinton would go on to defeat him in the November 1992 presidential elections. Yet George H.W. Bush's accomplishments in foreign policy, including the fall of the Soviet Union and the defeats of Nicaragua and Iraq, proved to be his lasting legacy.

By January 1992, 500,000 U.S. troops had left the Iraqi theatre. Yet Hussein would continue to defy U.N. weapons inspectors and wrongly kicked them out of Iraq. As a result, President Clinton and U.K. Prime Minister Tony Blair ordered a four-day strike on Iraq. Yet the United States could not go to Baghdad then to fight Hussein, because it would have lost Arab members of the coalition that it needed. But one day, in 2003, the U.S. military came knocking again for Hussein, this time on Iraq territory, in order to seek his capture. But it would take a horrific act of hatred on U.S. soil on September 11th, 2001 to make that happen.

Bibliography

1. *My American Journey*, by General Colin Powell, with Joesph E. Persico (Random House, 1995), disc 7.
2. *From Beirut to Jerusalem*, by Thomas Friedman (Farrar, Straus and Giroux, 1989), disc 3.

3. *Essential Histories: The Gulf War 1991*, by Alastair Finlan (Osprey Publishing, 2003), disc 7.

4. *George Bush: Life of a Lone Star Yankee*, by Herbert S. Parmet (previously listed).

5. *The Wars of America*, by Robert Leckie (previously listed).

6. *The Complete Book of U.S. Presidents*, by William A. Degregario (previously listed).

7. *A Critical Analysis of the Gulf War*, by Colonel Harry G. Summers, Jr. (Dell Publishing, 1992).

INTERVIEW WITH STEVE DAHL

Thomas Athridge: What was your rank and military branch?
Steve Dahl: E-5 in the U.S. Navy, Quartermaster Second Class, a navigator in the Navy.
TA: Where were you stationed, and what ship were you on?
SD: in San Diego, at the 32nd Street Naval Station, and onboard the guided missile cruiser the *U.S.S. Horne*, off the coast of Iraq in the first Gulf War. The CG-30 on the hull means cruiser-guided missiles.
TA: When did you start your military service?
SD: I served from 1989 to 1993, for four years.
TA: Did you agree with the first President Bush and his decision to launch military strikes against Iraq in 1991 to free Kuwait, and why?
SD: Yes, because I believe we had to go in to free the people of Kuwait, and the U.S. established a united coalition of nations to participate.
TA: Did you agree with Bush's decision to invade Panama and capture Noriega, and why?
SD: Yes, we did have every right to go after him, because he was attempting to block the Panama Canal. He was also running illegal drugs through his country.
TA: Should the U.S. military and its international coalition have entered Iraq in 1991 and toppled Hussein then, or was Bush right to stop at the borders of Kuwait?
SD: We should have continued on then into Iraq, because we would not have to face the war that we are in today if that had happened.
TA: Were you concerned about accidental deaths aboard your ship?
SD: No. We had one person die of a heart attack, but we were always training, and being prepared was very important.
TA: Did you worry about your ship being attacked, and why?
SD: Yes, because we trained for it often, and we did battle drills and damage control situations daily, such as battle stations and suiting up in fire gear. We prepared for chemical attacks, such as wearing gas masks and taking Atropine, which is medicine for a chemical attack. Biological weapons were a concern, also.
TA: Did the first President Bush successfully defeat America's "Vietnam syndrome," despite not being re-elected as president in 1992?
SD: Yeah, he broke it, because they were not thinking of Vietnam, but to try to help our allies and stabilize the Middle East region. I disagree with Bush's decision to pull out of the region with Hussein still left in power.
TA: Describe your experiences in the Navy during this war.
SD: We were at sea for 62 days until we pulled into Dubai in the United Arab Emirates, [and they] treated us well. We were in a minefield at sea

for 12 days, and did not realize it due to sonar! When we fired a battle gun from our ship, we moved back 25 feet. We diverted the Iraqis to think that we were going to do an amphibious landing in Kuwait all over, but we really went through Saudi Arabia. Two ships were hit with mines, the *Tripoli* and the *Princeton*, but there were no casualties and only aft steering damage. After the war, we went to Diego Garcia Island in the Indian Ocean for refueling, and then we went to Australia. We found a floating mine and we detonated it, and our ship pulled an Iraqi body out of the water.

TA: How do you feel about the second President Bush's handling of the Iraqi insurgency, and do you feel that a timetable for a pull-out is necessary?

SD: We have to end it and not leave the country in chaos, and we are doing a good job there, and we can come home when the job is done: to prevent another evil empire from springing up there. We should not bring the troops home on a timetable and should stay there until the job is done, no matter how long it takes.

Chapter 14
The War on Terror That Began on September 11th, 2001, and the Actions of President George W. Bush From 9/11 to Afghanistan to Iraq

The current war on terror is an issue that the United States has been dealing with, in regards to its relations with all the countries in the Middle East, since the late 1960s. This region of the world has often been involved in wars, beginning with the Israeli-Egypt War of 1967, which also involved both Syria and Jordan against Israel. The war lasted six days, and Israel captured Syria's Golan Heights, Jordan's West Bank, and Egypt's Sinai Peninsula. Another war with Israel followed in 1972, and Israel won that one, also. The Iranian hostage crisis in Iran began in 1979, when U.S. Embassy personnel in Tehran were taken hostage by the country's extremists in the new government of Ayatollah Khomeini and held against their will for 444 days. They were released on January 20th, 1981, which was the day that Ronald Reagan was sworn in as America's 40th president.

The Israeli-Palestinian conflict continued in 1982, when Palestinian militants attempted to take over the Lebanese city of Beirut; Israel sent its military to Beirut and captured the city. The Palestinian leader then was Yasser Arafat, who was elected Chairman of the P.L.O. (Palestinian Liberation Organization) in 1969 (pg 16, #2). In 1983, U.S. Marines were stationed in Beirut to protect the peace process there when, on October 23rd, 1983, a suicide car bomb made its way into the U.S. barracks and detonated a bomb, which murdered 241 Marines. This act of terror actually led President Reagan to withdraw U.S. troops from Beirut and bring them home. This decision by President Reagan was implemented because no one, from a U.S. military standpoint, had prepared for this kind of attack (pg 204, #2).

Indications at that point had suggested that both Syria and Iran were involved in the Marine bombing. When the Iran-Iraq War began in 1980, Reagan wanted Iraq to win, even though Iraq had invaded Iran first. In fact, former Secretary of Defense Donald Rumsfeld went to Iraq personally in 1983 to meet with Saddam Hussein and offer him U.S. support. Iraq and the United States had full diplomatic ties with each other then. Reagan strengthened Iraq militarily so it could efficiently fight Iran.

Also, Reagan strengthened Israel's military in an effort to counter Soviet influence in the region (pg 43, #2). Strong U.S. support for the nation of Israel gave Islamic radicals the ammunition needed to recruit terrorists to fight America. After the American military withdrawal from Beirut, Reagan invaded Grenada and showed that America could still win wars. Yet, during this time, tension between Iran and America was still high, which almost resulted in war between the two nations. The terror group Hezbollah and Iran were believed to be involved with each other.

When George H.W. Bush was inaugurated as president in 1989, he had to react to terror in the skies early in his presidency when a Libyan terrorist bomb detonated onboard Pan Am Flight 103 over Lockerbie, Scotland, on December 21, 1988, killing 259 innocent people. As a result, Bush imposed economic sanctions against Libya. With heightened tension between Iran and America, the *U.S.S. Vincennes* mistakenly shot down an Iranian passenger plane, wrongly believing it to be a fighter plane, killing 290 Iranian civilians (pg 101, #2). However, President George H.W. Bush did score some foreign policy successes with the ouster of Hussein's forces from Kuwait in 1991, and the capture of Manuel Noriega in Panama when he was accused of drug trafficking and harassing U.S. military personnel stationed in Panama, called "Operation Just Cause" (pg 691, #3). Noriega also nullified an election held in Panama that would have stripped him of power. U.S. forces landed in Panama to remove Noriega from power, and after four days of fighting Noriega was captured by the United States. Twenty-three U.S. troops were killed in this operation out of the 24,000 who participated, and 322 U.S. soldiers were wounded. Panama lost about 300 troops and 200 civilians, with an additional 100 troops and 2,000 civilians wounded (pg 692, #3).

In January 1993, William Clinton was inaugurated president after defeating President Bush in November 1992. Clinton and his Administration also had to deal with a fair share of international and domestic acts of terror against U.S. citizens and our allies. For example, in Bosnia, during 1993, the Serbs killed approximately 250,000 Bosnians, and an additional 2.5 million people were driven from their homes. Serbian leader Slobodan Milosevic, a former Communist leader, was to blame for this ethnic-cleansing atrocity. As a result, the U.S. created a no-fly zone over the country, which was once known as Yugoslavia. Just as bad, the situation in Somalia at the end of Bush's term and continuing into Clinton's was horrible, with a reported 350,000 Somalian people dying in civil war unrest in 1992 (pg 110, #4). Despite this, the U.S. military succeeded in carrying out its objectives of returning Somalia to a land of law and order.

On February 26th, 1993, a car bomb exploded inside the World Trade Center complex parking garage, which killed six people and wounded over

1,000. Terrorists Ramsi Yousef, Mohammed Salameh, and Sheikh Omar Abdel Rahman, a Sunni Muslim also known as the Blind Sheikh, were convicted in this crime. In April 1993, F.B.I. and A.T.F. (Bureau of Alcohol, Tobacco, Firearms and Explosives) agents raided the Branch Davidian compound in Waco, Texas, which resulted in the deaths of four agents and 76 Branch Davidian members, including women and children and their leader, David Koresh.

The U.S. military established a presence tin Bosnia to stem the carnage occurring there. Because of Serbia's aggression against the Croatian Muslims, led by Milosevic, the U.S. government imposed sanctions against it (pg 263, #4). Later, in October 1993, a U.S. humanitarian mission in Somalia went terribly wrong when U.S. soldiers on patrol were ambushed by Somalian rebels. This attack resulted in the deaths of 18 soldiers and the wounding of an additional 78. One soldier was even dragged dead through the streets of Somalia, and the film "Black Hawk Down" was based on this experience. To respond to this attack, President Clinton ordered more troops to respond to the violence in Somalia in order to stop it outright, and this plan was successful. Yet the bloodshed of the Somalian people in this situation was devastating and incredibly cruel.

On September 20th, 1994, President Clinton sent U.S. troops to Haiti in order to protect the rule of its democratically elected President, Jean-Bertrand Aristide. The Haitian guerilla rebels, determined to prevent Aristide's assumption of power, were overpowered by American forces. This mission was a military success for America. Later, on April 19th, 1995, the worst terror attack to date on American soil occurred when the Alfred P. Murrah Federal Building in Oklahoma City, Oklahoma was destroyed by a bomb: 168 people were killed in this evil attack, including women and children. More than 500 additional people were injured in this incident. Although this attack was initially blamed on international terrorists, it turned out that two Americans, Timothy McVeigh and Terry Nichols, who were responsible for this cowardly act of hatred. Nichols received a life sentence, and McVeigh was executed.

In a related story, Mir Aimal Kasi was sentenced to death and executed after being tried and convicted of killing two C.I.A. agents in Langley, Virginia in January 1993 (pg 411, #4). Despite this, the violent terrorism against American citizens continued. On June 25th, 1996, a truck bomb exploded at the U.S. Air Force base in Khobar, Saudi Arabia, at the Khobar Towers, resulting in the deaths of 19 U.S. servicemen. A Riyadh military building, in which the U.S. military trained Saudi military personnel, was also bombed, resulting in the deaths of five Americans. Terrorism continued to strike America at home when, at the Summer Olympics in Atlanta, Georgia, in 1996, a bomb went off in Olympic Park, despite extremely high levels of security for this event provided by the United States, including strategic placement of Black Hawk helicopters.

This public-square bombing in Olympic Park resulted in the death of one person, and injuries were sustained by an additional 14 people. Eric Rudolph was found to be responsible for this act, and he was captured in 2003.

On November 4th, 1995, the peace situation in the Middle East became compromised when Israeli Prime Minister Yitzhak Rabin was shot and killed by another Israeli who opposed the transfer of power of the West Bank to Palestine (pg 280, #4). Shimon Peres assumed power from the murdered Rabin, and fighting between the Israelis and Palestinians continued. A peace agreement was worked out in Bosnia between the Croats and Serbs, but only after the terrible price of 250,000 people murdered in this conflict. Al-Qaeda then would strike American interests after Osama bin Laden issued his fatwa in 1998, which resulted in the simultaneous U.S. Embassy bombings in Kenya and Tanzania on August 7th. These attacks resulted in the deaths of 224 people and injured 5000. As a result of these attacks, President Clinton declared war on terror (pg 129, #2).

As far back as September 9th, 1996, Clinton asked Congress for $1.097 billion for counter-terrorism activities, which he received (pg 130, #2). However, during 1998, Al-Qaeda merged with Egyptian Islamic Jihad in order to become stronger (pg 153, #2). While President Clinton wanted to declare all-out war on Al-Qaeda, he was hurt politically by the impeachment process being levied against him from House Republicans, stemming from an improper relationship with a White House intern. It was, however, on top of Clinton's list of national security objectives to take out Osama bin Laden and Al-Qaeda. For example, President Clinton signed Executive Order 13099, which imposed sanctions against Al-Qaeda and Osama bin Laden (pg 190, #2).

The only countries that had diplomatic relations with the Taliban government in Afghanistan were Pakistan, the United Arab Emirates, and Saudi Arabia. After all, only the lawless government of the Taliban would accept Osama bin Laden from Sudan after he was expelled from that country in 1996, and Osama bin Laden put his vast fortune to work in the Taliban-controlled country. Yet Pakistan, the United Arab Emirates , and Saudi Arabia pulled back diplomatic relations with the Taliban under heavy U.S. pressure to turn over Osama bin Laden to America. Then, towards the end of Clinton's presidency, in October 2000, the *U.S.S. Cole* that was stationed off the coast of Yemen was attacked by a suicide bomber on a small boat, killing 17 U.S. sailors and injuring 39.

Despite President Clinton being in office for some of the terror attacks in America and around the world, he did succeed in combating terror in several respects:

1.   Bosnia was successful in defeating Al-Qaeda.
2.   Once the troop massacre in Bosnia occurred, Clinton stepped up the U.S. military's presence there for six months, and, as a result, no more

U.S. troops died there. A U.N. coalition force replaced them after six months.

3.   America thwarted a plan by Al-Qaeda to simultaneously blow up eleven 747 airplanes over the Pacific Ocean by using liquid explosives.

4.   Clinton asked Congress for and received an increase of money to fight terror in the U.S. budget, from $5.7 billion in 1995 to $11.1 billion in 2000. Also, President Clinton's counter-terrorism Presidential Decision Directives in 1995 (no. 39) and May 1998 (no. 62) reiterated that terrorism was a national security problem and not just a law-enforcement issue (pg 108, #2).

After the embassy bombings in Africa during 1998, Clinton ordered U.S. missile strikes to be fired at Sudan and Afghanistan, but these strikes were not very damaging and did not stop the progression of Al-Qaeda. For example, in May 1999, Osama bin Laden could have been killed in Kandahar by U.S. cruise missiles, but America failed to take advantage of that opportunity (pg 140, #5). Clinton also did not have time to adequately respond to the *U.S.S. Cole* bombings, due to his leaving office, and thus halted a U.S. retaliatory attack. Clinton was very busy in his final days as president, trying desperately to solve the Israeli-Palestinian crisis, and to hurt terror organizations around the world. However, Clinton could not solve the Israeli-Palestinian crisis, despite his energetic efforts to secure a peace accord between the two nations. Also, Clinton was hesitant to bomb areas in Afghanistan without specific evidence of bin Laden's whereabouts, because he did not want to take any chances of killing innocent people in the air strikes, as occurred in the mistaken bombing of the Chinese Embassy in Belgrade, Serbia.

Osama bin Laden was an evil terrorist financier from Saudi Arabia, whose personal fortune was reported to be in the hundreds of millions of dollars. Osama bin Laden was originally involved with fighting the Soviets when they invaded Afghanistan in 1979. By the time the Soviets withdrew from Afghanistan in 1989, the U.S. and the rest of the world were guilty of sending no aid whatsoever to the Afghan people to help them reconstruct, which eventually led to the rise of the Taliban government there. After his participation in the Afghan-Soviet war had ended in 1989, Osama bin Laden began to contribute funds to radical Islamic groups located in the United States as early as 1991 (pg 66, #6). Osama bin Laden was outraged when American troops were stationed on Saudi Arabian soil in order to protect the country from invasion during the 1991 Desert Storm campaign, despite the U.S. military being asked to do so by the Saudi monarchy. Osama bin Laden wanted U.S. forces out of the Arabian peninsula, and also to topple the Saudi monarchy (pg 145, #6). Osama bin Laden was funding the first World Trade Center attackers in 1993 by providing living expenses while they were in America. One of the men responsible for the first World Trade Center attack in February 1993,

Ramzi Yousef, had an uncle named Khalid Sheikh Mohammed (pg 145, #5).

Khalid Sheikh Mohammed went to Chowan College in North Carolina, and later attended North Carolina A&T, where he graduated with a degree in mechanical engineering (pg 146, #5). Khalid Sheikh Mohammed broke with the United States over the Israeli issue. He went to work with bin Laden in Afghanistan in 1999. After the *U.S.S. Cole* bombing on October 12th, 2000, by a suicide team on a small boat, bin Laden had fled the scene for fear of an imminent American military attack, which never came. Yet before Clinton left office, his Administration was successful in stopping an Algerian terrorist named Ahmed Ressam, on December 14th, 1999, from driving from Canada to Los Angeles with explosives in a failed attempt to blow up Los Angeles International airport on New Year's Eve (pg 177, #5). President Clinton had declared war on terror in April 1996, and he desperately wanted to bring Osama bin Laden to justice before the end of his term in January 2001, but the United States had feared that a lack of evidence to try Osama bin Laden in a court of justice would preclude a conviction.

Osama bin Laden was originally from Saudi Arabia, but his fundamentalist views and connections to terror groups got him expelled from that country. In fact, Osama bin Laden was funding jihadists as early as 1992 (pg 109, #5). After Osama bin Laden's involvement in plotting the terror attacks at Riyadh and the Khobar Towers in Saudi Arabia, he was next expelled from Sudan on May 18th, 1996. Osama bin Laden next went to Afghanistan, which was the only country that would accept him, and he financed and trained an estimated 10,000 to 20,000 terrorist recruits in his training camps. Osama bin Laden issued his fatwa on February 23rd, 1998, which declared war on the United States. Although President Clinton wanted Osama bin Laden captured and brought to justice, he felt legal restraint from actually killing him due to the presidential ban on assassinating foreign leaders that was signed into law by President Ford (pg 6, #7). Clinton did order five military attacks to destroy Osama bin Laden, yet he escaped each time (pg 33, #7).

Osama bin Laden recruited oppressed and poor people to his terrorist training camps (#8). The rise of the Taliban and Al-Qaeda in Afghanistan can be blamed to some degree on America after the Afghan War with the Soviets ended in 1989, according to Richard Clarke:

1.   The C.I.A. was too dependent on Pakistani intelligence to aid the Afghan people, and as a result, America was not as involved as it should have been, considering the fact that the United States had helped the Afghans beat the Soviets. Once the Soviets left Afghanistan, so too did the Americans.

2.   The United States had no influence in the region because Congress cut budget money that was supposed to go to Afghanistan to save money.

3.   The United States did not help the Pakistanis deal with the flood of Afghan refugees and also cut financial aid to Pakistan.

4.   In the 1990s, the C.I.A. made the mistake of hiring the Afghans to kill Osama bin Laden instead of doing it themselves (pg 54, #2).

While Clinton wanted desperately to capture Osama bin Laden and defeat Al-Qaeda, in order to invade Afghanistan, the United States would have to go through Pakistan, which was a problem. In 1999, Pakistan was thought by the United States to back both the Taliban and Osama bin Laden, and it was very close to war with India (pg 124, #5). Consequently, Al-Qaeda and the Taliban took over Afghanistan's government in 1996. The Taliban was headed by Mullah Omar (pg 85, #8).

Clinton chose to fund the Northern Alliance in Afghanistan, led by Ahmad Shah Massoud. This idea was recommended by then-counter-terrorism leader Richard Clarke. In July 1999, Clinton had signed an Executive Order that declared the Taliban to be a state sponsor of terrorism. The United States would then impose economic and travel sanctions against Afghanistan in order to force Taliban leader Mullah Omar to give Osama bin Laden up to justice. But Omar refused to comply with this order. In response to this, both the United States and Russia persuaded the United Nations to pass U.N. Security Council Resolution 1333, which included an embargo on arms sales to the Taliban, in December 2000.

During this turbulent time in American foreign policy, the presidential election of 2000 was in disarray, with no official declared winner between then-Governor of Texas George W. Bush and then-Vice President Al Gore. This particular election for president was incredibly close. Though Gore had technically won the popular vote, a Supreme Court decision gave Florida's contested electoral votes for president to Bush, thus making him the 43rd President of the United States. Yet this result did not come for an unprecedented 36 days after the November 7th election. Because of this, Bush's transition team was hurt by the delay of the election results; terrorism was not really a campaign issue.

This meant that George W. Bush faced extra delays in getting his key appointments to his Cabinet confirmed by the Senate. During the transition of power from Clinton to Bush, Bush gave the C.I.A. power to kill Osama bin Laden after he was briefed on the subject. Clinton warned Bush that Osama bin Laden would be his biggest threat as president, and Clinton apologized to Bush for not successfully capturing or killing Osama bin Laden while he was in office. Yet Bush had different priorities than his predecessor, which included tax cuts, China, missile defense, the collapse of the Middle East peace process, and the Persian Gulf (pg 199, #5). However, both Bush and Clinton felt that Osama bin Laden's threat before the 9/11 attacks did not constitute sufficient evidence to warrant a ground invasion of Afghanistan.

Meanwhile, in the Taliban-controlled Afghanistan, Ahmad Massoud wanted to undertake a major offensive against the Taliban. Massoud and his Northern Alliance forces, at the time, only controlled a small portion of Afghanistan. Afghanistan's mountains dominate the land area. The terrain makes up some of the world's most rough and harsh regions. There are a reported 10 million land mines throughout the nation of Afghanistan. Warfare has been the country's entire history. In fact, the most common form of death is assassination, or combat death. Massoud had witnessed this brutality in Afghanistan for 20 years. At the time, Afghanistan had the largest amount of refugees in the world, due to being driven out by the Taliban. Also, millions of Afghans (#2) were killed in the almost ten-year war with the Soviets (pg 82, #8).

Massoud did well militarily in a place called the Panjshir Valley against his Taliban foes. Massoud's vision for a new Afghanistan was that it would be completely democratic, including protecting rights and education for women, and remain a Muslim-dominated country. Massoud continued to fight the Taliban and Osama bin Laden with his Northern Alliance soldiers, despite receiving very little help from other countries.

Massoud's message to President George W. Bush in April 2001, at a conference in Paris, France, was that terrorist attacks against the United States were coming. To support this, U.S. Ambassador of Indonesia Robert Gelbard warned of the impending terror threat from Al-Qaeda as early as 2000. Moreover, an EgyptAir flight from New York to Cairo, Egypt had plunged into the Atlantic Ocean when the Egyptian co-pilot intentionally crashed the airliner, which killed 213 innocent people (pg 277, #6). This act may have inspired Al-Qaeda to use commercial airliners to destroy ground targets. Richard Clarke served both the Clinton Administration and the Bush Administration as the counter-terrorism czar (pg 200, #5).

On January 25th, 2001, Clarke had written a letter to National Security Advisor Condoleezza Rice, in which he urgently requested the Bush Administration to deal with Al-Qaeda, but the official meeting on the subject did not occur until right before 9/11. Meanwhile, Massoud again warned the United States that unspeakable terrorist actions were coming soon. This statement from Massoud was echoed by Clarke on September 4, 2001, when he told Rice that an attack on the United States from Al-Qaeda was definitely coming. Despite this, then-Attorney General John Ashcroft had turned down an offer to increase the Department of Justice counter-terrorism budget by $58 million on September 10th (pg 298, #6). But by then, the momentum leading to the events of the horrible tragedy that was 9/11 was already well under way.

On September 9th, 2001, in order to have him out of the picture, two Al-Qaeda members who were posing as journalists killed Massoud and themselves in Afghanistan in a suicide bombing. With Massoud eliminated, the Al-Qaeda hijackers were able to carry out their mission of evil. All of the nineteen 9/11 hijackers bought their airline tickets between August 25th and September 5th, 2001 (pg 249, #5). When Al-Qaeda terrorist Zacarias Moussaoui was arrested in August 2001, for suspiciously asking how to take over an aircraft and fly the plane once already in the air, there was a tragic miscommunication between the Minneapolis police officers and the F.B.I. over the role that Moussaoui played in the overall plot of terror planned for 9/11.

While in custody in Minnesota, Moussaoui confessed to the Minneapolis officers that there were plans to crash hijacked planes into the World Trade Center. However, when this information was reported to the F.B.I., the agency did not believe it (pg 275, #5). In fact, the F.B.I. did not even know that Moussaoui was an Al-Qaeda member until after the 9/11 tragedy, when the London office informed it of his training in Afghanistan. In review, had the Minneapolis police been successful in convincing the F.B.I. to search Moussaoui's computer, it might have been enough to have prevented the 9/11 tragedy, but we will never know. The reason for this denial of a search by the F.B.I. of Moussaoui's computer was a lack of probable cause.

The rest of the nineteen evil-doers did succeed in entering the United States legally, and learned to fly airplanes at flight schools. They trained for this mission for two years (pg 23, #8). Al-Qaeda used mostly Saudis to carry out their evil mission, because they would be less suspicious and would fit into American society better. Even though some of the nineteen hijackers were on terror watch lists and warnings, they purchased their tickets with relative ease and boarded their flights on that horrible day without significant problems.

Moussaoui's C.I.A. case file at the time of 9/11 was called "Islamic Extremist Learns to Fly." The terrorism and evil that Osama bin Laden and Al-Qaeda thrived on was different from anything that the U.S. government had faced before, specifically because no terror group was responsible for acts of destruction without an affiliated country. Even though Afghanistan harbored these terrorist training camps, Clinton and the second Bush Administration felt that Osama bin Laden's threat before 9/11 was not enough to warrant a U.S. ground invasion of Afghanistan (pg 350, #5). However, according to the 9/11 Commission report, there were 10 missed operational opportunities by the U.S. government that could have prevented 9/11:

1.   January 2000: the C.I.A. did not put Khalid al-Mihdhar on its watch list or notify the F.B.I. when it learned that Mihdhar possessed a valid U.S. visa.

2.   January 2000: the C.I.A. did not develop a transitional plan for tracking Mihdhar and his associates so that they could be followed to Bangkok and onward to the United States.

3.   March 2000: the C.I.A. did not put Nawaf al-Hazmi on its watch list or notify the F.B.I. when it learned that he possessed a U.S. visa and had flown to Los Angeles on January 15th, 2000.

4.   January 2001: the C.I.A. did not inform the F.B.I. that a source had identified Tawfiq bin Attash, who was a major figure in the October 2000 bombing of the *U.S.S. Cole*, as having attended the meeting in Kuala Lumpur with al-Mihdhar.

5.   May 2001: a C.I.A. official did not notify the F.B.I. about al-Mihdhar's U.S. visa, Hazmi's U.S. travel, or al-Mihdhar's having attended the Kuala Lumpur meeting.

6.   June 2001: F.B.I. and the C.I.A. officials did not ensure that all relevant information regarding the Kuala Lumpur meeting was shared with the *Cole* investigators at the June 11th, 2001 meeting.

7.   August 2001: the F.B.I. did not recognize the significance of the information regarding al-Mihdhar and Hazmi's possible arrival in the United States, and thus did not take adequate action to share information, assign resources, and give sufficient priority to the search.

8.   August 2001: the F.B.I. headquarters did not recognize the significance of the information regarding Moussaoui's training and beliefs, and thus did not take adequate action to share information, involve higher-level officials across agencies, obtain information regarding Moussaoui's ties to Al-Qaeda, and assign sufficient priority to determining what Moussaoui might be planning.

9.   August 2001: the C.I.A. did not focus on information that Khalid Sheikh Mohammed was a key Al-Qaeda lieutenant, or connect information identifying him as the "Mukhtar" in other reports to the analysis regarding Ramzi bin al-Shibh and Moussaoui.

10. August 2001: the C.I.A. and the F.B.I. did not connect the presence of al-Mihdhar, Hazmi, and Moussaoui to the general threat reporting about imminent attacks (pgs 355-56, #5).

By September 11th, 2001, Osama bin Laden had successfully trained 10,000 to 20,000 recruits in his terrorist training camps (pg 67, #5). bin Laden's extremist views on Islam and its history is designed to appeal to the Sunni sect and Arabs. Sunnis believe that lineal descent of the prophet Mohammed was not a requirement of leadership (Ummah), but the Shias believed that the leader of the Ummah must be a direct descendant. In fact, a classified report was on President Bush's desk about the Al-Qaeda organization and its capabilities on that fateful morning, but because the president was in Florida on an education reform visit, he did not have time to read its contents. He was sitting in front of a group of schoolchildren on the morning of September 11th, 2001 when the terrorist attacks and subsequent tragedies were taking place, and the president was informed of the first plane hitting the World Trade Center by White House Chief of Staff Andrew Card. At first, there was speculation that the airliner hitting the building was just a horrific accident, but Card informed the president of the second plane hitting the other tower 20 minutes later, saying, "A second plane hit the second tower. America is under attack" (pg 38, #5). Bush knew that this nation had been ruthlessly attacked by a group of cowards and that the United States, as a result, was at war with the terrorists.

Bush's quote to the media directly after the terror attacks was, "Terrorism against our nation will not stand." This statement was similar to his father's statement when he was president, "This will not stand," which was in reference to the invasion of Kuwait by Iraq (pg 16, #7). Despite the horror that had already taken place that awful morning, it would continue when, at 9:39 a.m., American Airlines Flight 77 was deliberately crashed into the Pentagon, which resulted in 189 innocent people being killed. All 64 people aboard the airliner were killed, along with 125 people inside the Pentagon (pg 314, #5). Defense Secretary Donald Rumsfeld was in the Pentagon at the time of the crash. Meanwhile, the horrific scene at the World Trade Center was only beginning. There were an estimated 1,366 people who were trapped above the impact site at Tower One, and none of them survived. About 200 of them were forced to jump from the buildings to their deaths. Tragically, the Port Authority was powerless to rescue the people who were trapped above the fire, because it had no protocol to do so. Also, helicopters could not land on the roof of either tower, because it was not safe for the pilots to land.

Since the difficulties of people escaping from the World Trade Center during the first bombing that occurred there in February 1993, the New York Fire Department and the Port Authority developed improvements in ways to escape form the buildings safely. They had established "fire-safety teams" on each floor of the World Trade Center, which did help save people's lives during the 9/11 attacks on the buildings (pg 281, #5). However, this was one of the worst disasters in American history, one that was manmade. Despite this enormous catastrophe inflicted on both World Trade Center buildings, absolutely no one thought that either tower would collapse (pg 304, #5). But by 10:30 a.m. on that awful morning, that is exactly what had happened.

In the 9/11 attacks on New York, Washington, and Shanksville, Pennsylvania, 2,973 innocent people were murdered, the single largest loss of life on American soil as the result of hostile aggression in its history, including 343 New York Fire Department deaths. After the Pentagon crash that terrible morning, a fourth plane was hijacked, United Airlines Flight 93, flying from Newark, New Jersey, to San Francisco, California, that was intended to strike the Washington, D.C. area a second time, at either the White House or the Capitol. But brave passengers fought back against their attackers and forced the hijackers to crash the plane into a field in Shanksville, Pennsylvania at 10:06 a.m., which resulted in the deaths of the 40 passengers and crew and their four hijackers. Richard Clarke's warning to Rice on September 4th, 2001 proved to be true.

When President Bush was informed of the magnitude of the attacks of September 11th, 2001, his statement was, "Someone is going to pay" (pg 17, #7). He also said, "Make no mistake, the United States will hunt down and punish those responsible for these cowardly acts" (pg 16, #2). Immediately after the attacks on that day, and for precautionary measures, the president went from Florida to Barksdale Air Force Base in Louisiana, to Strategic Command headquarters in Omaha, Nebraska, in order to confuse the terrorists. The president returned to the White House later that night to address the stunned nation from the Oval Office, in which he said, "Freedom itself was attacked this morning by a faceless coward, and freedom will be defended" (pg 17, #2).

Later that evening, when President Bush was meeting with his "War Cabinet," which consisted of National Security Advisor Condoleezza Rice, Chief of Staff Andrew Card, Vice President Dick Cheney, Secretary of Defense Donald Rumsfeld, Attorney General John Ashcroft, F.B.I. Director Robert Mueller, C.I.A. Director George Tenet, and Chairman of the Joint Chiefs of Staff General Hugh Shelton, he formulated the plan of the U.S. military punishing not only the perpetrators of the attacks, but also the nations that harbored them. Bush stressed that the United States was now at war with a different kind of enemy. The case was then handed to Defense Secretary Donald Rumsfeld. By order of command in defense, President Bush was first, then Defense Secretary Rumsfeld, then General Tommy Franks at U.S. Central Command.

Rumsfeld was originally President Gerald Ford's Chief of Staff, and then moved over to Defense Secretary for Ford when Dick Cheney became Chief of Staff. Rumsfeld recommended that the United States use military, legal, financial, and diplomatic resources, including the C.I.A., to root out and defeat terrorism (pg 33, #7).

It would take commercial airline flights three days to resume, under a reduced, more supervised schedule. Bush did return to Washington, D.C. on the night of September 11th, 2001, despite being advised against doing so by some Administration officials (pg 28, #7). Bush wanted to form an international coalition between the countries of Pakistan, Russia, China, and the United States for an attack on Al-Qaeda in Afghanistan. However, both Rumsfeld and Deputy Defense Secretary Paul Wolfowitz recommended to Bush that he consider also striking Iraq in the immediate aftermath of 9/11. To compound the problem, Al-Qaeda was operating in over 60 countries. Bush recorded in his personal diary the night of September 11th, 2001, "The Pearl Harbor of the 21st century took place today" (pg 37, #7).

When President Clinton was in power, he had issued five military orders to kill bin Laden, but the terror mastermind escaped each time. Now, Bush chose to aid both financially and militarily the Northern

Alliance in order to fight the Taliban in Afghanistan. The Taliban outnumbered the Northern Alliance two to one (pg 35, #7). Bush was told on September 12th, 2001 by C.I.A. Director George Tenet that the previous day's attacks had taken two years to plan (pg 39, #7). Based on that assessment, Bush said that combating terror would be his main focus for the remainder of his term (pg 41, #7). Bush described the enemy as people who operate their business in the shadows, prey on innocent people, and then run for cover (pg 45, #7). President Bush classified the struggle as "good versus evil."

While Bush had the support of Great Britain, he also wanted to form a coalition to fight terror with France, Germany, China, and Canada. However, if need be, Bush was prepared to have the United States go to war alone (pg 45, #7). General Tommy Franks was in charge of the South Asia and Middle East divisions of the U.S. military, and was therefore in charge of planning the U.S. invasion of Afghanistan (pg 43, #7). The United States demanded that the Taliban government in Afghanistan produce bin Laden and his deputies and chief lieutenants, including Ayman al-Zawahiri, and shut down all the terrorist training camps, or it would destroy the terrorist infrastructure. At the time, there was no direct link between the September 11th attacks and Hussein's Iraq, and bin Laden resented the secularism of Hussein's Iraqi regime. However, there was broad Administration support for striking Iraq, and Rice noted that the Administration was concerned that Iraq would take advantage of the 9/11 attacks to further promote its aggression. Deputy Defense Secretary Paul Wolfowitz also strongly argued to strike Iraq during "this round" of the war versus terror.

However, at this stage, Bush saw Afghanistan as the first priority. This decision by Bush was supported by both Secretary of State Colin Powell and Deputy Secretary of State Richard Armitage. Powell and Armitage wanted to obtain several agreements from Pakistan for its cooperation in the U.S. war against terrorism. Pakistan was led by General Musharraf. The seven stipulations were:

1.   Stop all Al-Qaeda operatives at Pakistan's border, intercept arms shipments through Pakistan, and end all support for bin Laden.
2.   Blanket overflight and landing rights.
3.   Access to naval and air bases in Pakistan for U.S. use.
4.   Immediate intelligence and immigration services.
5.   A complete condemnation by the Pakistani government of the 9/11 attacks on the United States.
6.   Cut off fuel supplies to the Taliban.
7.   Completely break relations with the Taliban government.

The United States got everything that they asked for from Pakistan, and this act produced stronger relations between the two countries (pg 59, #7). Bush did not want to upset the Pakistani people, who might rebel against Musharraf. Pakistan, like India, does possess nuclear weapons.

Before he could approve the plan of war for the United States, Bush had to first console the nation during this extreme time of grief for the thousands of families who were personally affected in the immediate aftermath of this awful tragedy.

Bush visited the World Trade Center ruins a couple of days after the tragedy, a scene that he would describe as "a nightmare, a living nightmare" (pg 69, #7). Rescue workers at the demolished World Trade Center site wanted revenge against the perpetrators who committed this cowardly attack. When Bush spoke to the rescue workers, he had to speak through a bull horn, which was lent to him by one of the workers. President Bush vowed to the rescue workers that he would find those who were responsible for this horrific attack. These criminals would be prosecuted to the fullest extent of the law (pg 70, #7). Bush also visited with family members of the victims at the Jacob Javits Convention Center (pg 71, #7).

When Bush returned to Camp David after his New York City visit, he met with his senior advisers (Cheney, Rice, Powell, and Rumsfeld) and constructed a plan of attack to root out the Taliban regime and bin Laden from Afghanistan (pg 72, #7). President Bush approved the American military plans to invade Afghanistan and take out the Taliban government on September 21st, 2001. On October 2nd, Central Command General Tommy Franks and his military team and advisers constructed the formula that was originally titled Operation Infinite Justice, but was changed to Enduring Freedom, in an effort not to offend the Muslim community. There were three phases of military operation in this plan:

Phase One: the U.S. and its allies would move forces into the region and arrange to operate from or over the neighboring countries to Afghanistan, such as Pakistan and Uzbekistan.

Phase Two: air strikes and Special Operations attacks would take out important Al-Qaeda and Taliban targets. The C.I.A. and the Special Operations forces would be deployed to work together with each major Afghan faction that was opposed to the Taliban. The raids were scheduled to begin on October 7th, 2001.

Phase Three: the United States would carry out "decisive operations" using all elements of national power, including ground troops, to topple the Taliban regime and eliminate Al-Qaeda's sanctuary in Afghanistan. Mazar-e-Sharif, in northern Afghanistan, fell to a coalition assault by Afghan and U.S. forces on November 9th. Four days later, the Taliban had fled from Kabul. By Early December, all major cities had fallen to the coalition.

Bush empowered the C.I.A. with the most broad and lethal authority in its history to combat our new 21st-century enemy. The Al-Qaeda organization is also located heavily in Indonesia, the Philippines, Yemen, Malaysia, and Somalia (pg 90, #7). Deputy Defense Secretary Wolfowitz said to National Security Advisor Rice that there was a 50% to 100%

chance that Iraq was involved with the 9/11 attacks, but he was proven incorrect. However, the Cabinet unanimously voted in favor of the U.S. military striking Afghanistan first. But it was also agreed that the United States must go after Saddam Hussein eventually.

In order to prevent future terrorist attacks against America, including the anthrax attacks, which were responsible for killing five people in late 2001, President Bush created from scratch the Department of Homeland Security. Homeland Security was to involve not just country and airline security, but also the prevention of biological and chemical attacks against our people, such as the aforementioned anthrax attacks. By creating the Department of Homeland Security, Bush realized that all agencies that involve military and law enforcement must come together under the branch of one federal department in order to properly fight the enemy (pg 93, #7).

Accordingly, Bush opened two fronts in the new 21st-century war on terror:

1. In Afghanistan
2. At home with homeland security

However, Bush made it clear that once the Taliban had been conquered and bin Laden either captured or killed, the Republic of Iraq was next on Bush's hit list. On September 24th, 2001, Bush signed an Executive Order to attack financial assets of global terrorists. This was specifically done to freeze assets of terror suspects. These suspects were placed on a terrorism watch list. Our strategy to dismantle the Al-Qaeda network was to attack its terrorist cells and their organization, prevent the continuing growth of Islamic terrorism, and prevent and prepare for terrorist attacks. Yet the war is not directed against Islam the religion, which is not synonymous with terror. Islamic texts do not condone terror in any form, and such a false equivalence misrepresents a peaceful religion that many people around the world believe in.

Vice President Cheney was most concerned about terrorists acquiring weapons of mass destruction and drug labs. Surprisingly, despite Vice President Cheney being heavily involved with U.S. policy making during the height of the Cold War with the Soviet Union, he recommended to President Bush that the United States ally with Russia for the Afghanistan invasion (pg 129, #7). Bush wanted food supplies to be dropped to the Afghan people in order to show the U.S. resolve to keep them terror free. Bush wanted to be seen as a liberator and viewed these humanitarian acts in Afghanistan as a morale issue.

National Security Advisor Rice inquired about Bush's strategy against other countries hostile to the United States, such as Iran, Iraq, Libya, Syria, and Sudan, because they had supported terrorist acts against our nation in the past (pg 131, #7). The Bush Administration then became obsessed with the prevention of terror on American soil. For example, Bush received a report called "Trying to Anticipate the Next Attack" on

September 25ᵗʰ, 2001, in which the U.S. intelligence agencies tried aggressively to think like the terrorists themselves (pg 132, #7). The C.I.A. then called together its own investigative team, called the "Red Cell," for the purpose of tracking terrorist actions. This investigation included nine categories:

1.   Political centers, such as Washington, D.C. or federal offices anywhere

2.   Infrastructure facilities, such as airports, roads, harbors, railroads, dams, tunnels, and bridges

3.   Economic systems, including Wall Street or Chicago trading centers

4.   Energy infrastructure, including refineries and oil platforms

5.   Military targets, such as areas of large troop concentrations (Army, Navy, Air Force, or Marine bases) or weapon storage sites

6.   Global telecommunication systems, such as electric communication transition points and internet centers

7.   Educational centers like Harvard or M.I.T.

8.   Cultural centers like Hollywood or sports stadiums, particularly during concerts or games

9.   Historical monuments and other symbols of national identity (pgs 132-33, #7).

The Bush Administration and the C.I.A. saw that these places and sites in America had to be strengthened with more security and better intelligence. Bush thought of the Pearl Harbor attack scenario replaying itself on September 11ᵗʰ, 2001, thus forcing the nation into war. Bush said that he must respond in the same way that President Franklin Roosevelt did (pg 91, #14). Afghanistan was the terrorist training site and planning area for the 9/11 attacks upon the United States, as well as the haven of bin Laden, the number one most-wanted terrorist in the world, and the U.S. military was to strike the first blow in the war against terrorism there (pg 369, #5).

President Bush was heavily influenced by Vice President Cheney, who stated that the number one biggest threat to national security in the United States was from terrorists acquiring weapons of mass destruction, and this extreme threat would exist for generations to come (pg 132, #14). The United States put massive pressure on the Taliban government to turn bin Laden and the other Al-Qaeda terrorists over to justice, but the Taliban refused to do so. Consequently, on October 7ᵗʰ, 2001, the U.S. military began its offensive against the Taliban and Al-Qaeda. The Al-Qaeda soldiers consisted of Pakistanis, Chechens, Arabs, and Egyptians (pg 17, #9). The U.S. military firepower in this offensive greatly overmatched that of its enemies. For example, an American AC-130 Spectre gunship can cover an entire battlefield with more rounds of ammunition than any other combat platform in the world, and many of them were used in the war in Afghanistan. The AC-130 Spectre gunship is nicknamed "raining death" (pg 74, #9).

The Taliban fighters were forced to surrender because they could not match the overall power of the U.S. military (pg 81, #9). The U.S. and Northern Alliance forces used Daisy Cutter high-powered explosives when they attacked the Afghan city of Mazar-e-Sharif, which left a 600-yard radius of devastation (pg 301, #7). The city fell to the U.S./Northern Alliance forces on November 9th, 2001. These two forces also liberated the Afghan cities of Taloquan and Kunduz. Green Beret forces for America were deployed in Afghanistan because they were all experts at unconventional warfare. In fact, on November 11th, 2001, the Special Forces A-Team called Triple Nickel launched 25 air strikes on Al-Qaeda, which resulted in 2,200 enemy casualties and destroyed twenty-nine tanks and six command posts (pg 309, #7).

The last stronghold for the Taliban in Afghanistan was Kunduz, which was captured on November 23rd, 2001 (pg 153, #9). This was where U.S. forces actually captured an American Al-Qaeda member named John Walker Lindh. Special Forces and the coalition military next captured Tarin Kowt, and went on to Kandahar, which is located in southern Afghanistan. After some intense fighting, the city of Kandahar was in American/Northern Alliance control by December 7th, 2001. Mullah Omar's Taliban government in Afghanistan completely fell by December 9th, 2001. Despite this, thousands of Al-Qaeda members escaped, including Omar and bin Laden (pg 242, #9).

The last major pocket of Al-Qaeda resistance in Afghanistan was Tora Bora (#10). The Eastern Alliance commander was Haji Zaman Ghamsharik, who believed that several hundred Al-Qaeda members fled from their caves into Pakistan, numbers that included an estimated two thirds of the original Chechen, Arab, and Afghan members of Al-Qaeda (#10). Osama bin Laden was reported to have remained in the Tora Bora region until the very end of the battle, which for him was about or around December 16th. He then managed to successfully escape Tora Bora and fled into the rugged Afghan/Pakistani mountain range. Tora Bora is in eastern Afghanistan, about 35 miles southwest of Jalalabad, and is described as a fortress of snow-capped peaks, steep valleys, and fortified caves. This area was used for the mujahideen during the Soviet-Afghan war. Osama bin Laden knew the caves well, having spent a decade of his life there in that war ("Lost at Tora Bora," *New York Times Magazine*, September 15th, 2005). Despite this fact, Brigadier General James N. Mattis and his 4,000 Marine troops could have sealed off all of the escape routes of Al-Qaeda fighters, and even Osama bin Laden himself, but were not allowed to carry out this action due to the heavy reliance on local Afghan fighters to bring bin Laden to justice. This act is described as the biggest mistake of the war by the U.S. military. General Franks had given this order from US Central Command headquarters in Tampa, Florida.

While the Battle for Tora Bora was a victory for U.S. and Northern Alliance forces, it was unsuccessful in rounding up all the remaining Al-Qaeda members and leaders, who likely escaped into the Pakistani/Afghan mountains.

Osama bin Laden had reportedly many people to assist him in his escape, including an underground railway maintained by sympathetic Afghan families (#10). He also used massive amounts of cash to help him escape. The Pakistani government did not allow U.S. troops to cross its border in search of Osama bin Laden and other Al-Qaeda members, but insisted that it would hunt down these criminals without outside help. Osama bin Laden was also known as "The Sheikh" (pg 113, #12). The warlords in Jalalabad, who were relied on to capture and bring down Osama bin Laden and Al-Qaeda, were proven to be unreliable. This area at the Afghan/Pakistani border, known as the White Mountains, has many caves, and not enough U.S./coalition troops to cover them all. To add to the frustration, American intelligence from both inside and outside Tora Bora was inaccurate (#10).

The Bush Administration blamed the escape of Osama bin Laden and other Al-Qaeda fighters from Tora Bora on not having enough U.S. troops on the ground to properly round up the evil-doers, as Afghan warlords let them escape. After all, there were more officers on the ground in D.C. after the sniper shootings in October 2002 than were in Tora Bora chasing after Osama bin Laden (pg 134, #12). Despite this, the Taliban government was officially ousted from Afghanistan, and Hamid Karzai was officially elected President of Afghanistan by the Afghan people, who had seen years of brutality and injustice occurring in their land.

The biggest concentration of firepower by the U.S./Northern Alliance to date would become known as "Operation Anaconda," in the Shah-i-Kot Valley in February 2002. The Al-Qaeda fighters in the Anaconda battle were mostly Arabs and Chechens (pg 188, #12). On the coalition side of the battle, 900-plus U.S. soldiers participated, along with 900 mujahideen soldiers and hundreds of soldiers from Australia, Canada, France, Germany, Norway, and Denmark. Also, the 101st Airborne Division's AH-64 Apache ships arrived to pound Al-Qaeda positions (pg 276, #9). Despite Operation Anaconda's success in forcing Al-Qaeda from the Shah-i-Kot Valley, the Al-Qaeda fighters had the advantage of knowing the caves , which is how they escaped (pgs 293-94, #9). The U.S. military lost eight men in Operation Anaconda, with an additional thirty-four wounded (pg 191, #12). This battle lasted until March 18, 2002.

There were more U.S. military personnel at the Anaconda battle than at Tora Bora (pg 200, #12). It is possible that if the Green Beret forces who were there, led by Colonel John Mulholland, had mixed properly with the conventional U.S. Army at Tora Bora, this could have led to far more Al-Qaeda captures and killings, including Osama bin Laden himself. This act could have dealt a knock-out blow to the Al-Qaeda organization, but it never happened. Author Phillip Smucker said that this fact should not be blamed on military leaders, such as General Tommy Franks or Colonel John Mulholland, but on the civilian government and the politicians at home (#12). Mulholland's Special Operations Mission in Afghanistan was called "Operation Task Force Dagger." Mulholland declared Operation Task Force Dagger a success in March 2002.

A high-ranking Al-Qaeda member, Khalid Sheikh Mohammed, was captured in Pakistan. While the U.S. and its allies were unsuccessful in finding Osama bin Laden, they were successful in finding two million rounds of ammunition, including mortars, rockets, and tank rounds, that were hidden in various caves. The Taliban government, and the evil that it represented, was terminated as the power in the country of Afghanistan. This war of terror that the United States was and now is in became a war to systematically destroy sophisticated organizations of terror. The next target on America's list of terrorist states returned us to a country with whom the U.S. and its allies had previously fought a war: Iraq.

Only ten days into President George W. Bush's presidency, he, his Cabinet, and his Administration were already looking to invade Iraq (pg 75, #13). When President George H.W. Bush declared war on Iraq in 1991, he and then-Secretary of State James Baker had a coalition of over 100 nations. However, President George W. Bush was prepared to go to war with Iraq with a much smaller coalition of nations. Defense Secretary Donald Rumsfeld pointed out to President Bush and the Cabinet that Iraq had more military targets to hit than Afghanistan did. Two countries in particular were at odds with the U.S./U.K. decision to go to war with Iraq, and they were France and Russia. This was because these two countries had invested heavily in business interests there (pg 15, #14). Defense Secretary Rumsfeld recommended to the president to go after Iraq literally hours after the attacks of September 11th, 2001.

Hussein was responsible for expelling all U.N. inspectors from Iraq in 1998 (pg 39, #14). Two men who had initially countered the Iraqi invasion idea were Secretary of State Colin Powell and Deputy Secretary of State Rich Armitage. These two men believed that the United States should strike Iraq without building a united coalition of nations first, as had been previously done by President George H.W. Bush's Administration in 1991 (pg 39, #14). Contrary to the Secretary of State's views, however, Bush knew that he wanted Hussein toppled from power and brought to justice as early as the end of 2001 (pg 52, #14).

The George W. Bush Administration's top priority was preventing terror organizations, or nations who might harbor them, namely Iraq, from acquiring weapons of mass destruction. Consequently, the Bush policy on combating terror was to strike it before it could strike the United States again. The face of terror became that of Saddam Hussein. The case for war against Hussein and Iraq was launched because U.N. inspectors were kicked out of Iraq in their search for weapons of mass destruction in 1998.

In order to understand the reasons and events that led to the war in Iraq, it is important to look back at the history of the republic, and what events led it to its present conditions. The country of Iraq is home to two great rivers, the Euphrates and the Tigris (pg 2, #15). Iraq was once called Mesopotamia (pg 8, #15). Islam became the dominant force in the region in the 8th century A.D. (pg 9, #15). The Ottoman rulers of that time divided Iraq into three governorships: Mosul in the Kurdish north, Baghdad in the Sunni center, and Basra in the Shiite south (pg 9, #15).

By the end of World War One in 1918, Iraq was controlled by Great Britain, who had beaten the Ottomans in that war. Of course, Austria-Hungary and Germany were also beaten in that war. At that time, Iraq was democratic, with an elected parliament. Also, Iraq was admitted into the ill-fated League of Nations as a sovereign and independent state. Later, in 1948, most of Iraq's 100,000 Jewish citizens moved to the newly created Israel (pg 23, #15). Hussein was a Ba'athist who was exiled from Iraq to Egypt in the late 1950s and early 1960s. The Ba'ath Party of Iraq consisted of Sunni Muslims. During his time in exile, Hussein was technically an anti-Communist Arab Nationalist.

Hussein's evil mind combined elements of Socialism, Fascism, and Pan-American Nationalism that he developed as a youth. Hussein was described during his younger years as a bully and a loner. He had tried in 1968 to kill then-Iraqi Prime Minister Abdul Karim Kassim, but failed in his attempt (pg 4, #16). After Hussein was released from jail and had returned to Iraq from exile, he became vice president in a successful Iraqi coup led by Ahmed Hassan al-Bakr, who became president. Hussein eventually seized power from the ailing al-Bakr and took full control of Iraq on July 22nd, 1979, at a press conference in Baghdad in front of 1,000 senior members of the Ba'ath Party. He then had the names of his 65 political adversaries read out by his henchmen, and had them escorted outside, where they were executed. Hussein called these horrible acts of terror "democratic executions." He videotaped these executions and made the Ba'athists control everyone and everything in Iraq. He modeled his reign after his hero, Joseph Stalin, and modeled his power on the Soviet control of the region.

Hussein then made all teachers join his party, along with journalists, writers, and artists, who were forced to join Iraq's General Confederation of Academicians and Writers. Hussein did initially allow the Iraqi court system to remain intact, but the Baghdad Revolutionary Court took control of the more important cases, in which no appeals were allowed. In Hussein's first year of rule, the Iraqi Popular Army doubled in size to 250,000 members (pg 53, #18). Trade unions or different party systems were not allowed, and were punishable by death. Hussein has been described as "a monster of cruelty and aggression" (pg 55, #15).

Ironically, Hussein was never a soldier. In fact, Hussein could not enter the Iraqi Military Academy in Baghdad as a youth, due to his lack of education. When Hussein took control of Iraq, he immediately declared war on Iran and attacked that country first. This war proved to be fruitless, and dragged on for years, until 1988. By then, Iraq's Army had sustained one million casualties. Also, the Iraqi war debt was approximately $80 billion, and reconstruction costs were estimated to be approximately $230 billion.

Hussein and Ayatollah Khomeini clashed because Khomeini was an Islamic fanatic who represented the Shi'ites, as opposed to Hussein's Sunni representation. They strongly differed in ideas. Hussein used chemical weapons against both the Iranians and his own people (the Kurdish population of Iraq), who became the first people since the Holocaust to be gassed by their own government (pg 13, #16). It is estimated that Hussein was responsible for the deaths of between 100,000 and 200,000 people. These acts were violations of the laws of warfare as stipulated by the Geneva Conventions, which alerted the United Nations of Iraq's crimes. This terror operation was assisted by Hussein's cousin, Ali Hassan al-Majid, or "Chemical Ali" (pg 60, #15).

Despite these acts of human cruelty, foreign countries continued to supply Iraq with weapons, including Egypt, France, the Soviet Union, and the United States. However, once Hussein invaded Kuwait in August 1990 in order to take its oil reserves, and threatened to strike oil-rich Saudi Arabia next, the U.S. military and its allies around the globe successfully forced Hussein and his army from the country of Kuwait. Twelve years later, with Hussein still in power, the U.S. military returned to Iraq, and the next phase of the war versus terror began.

Even though Hussein and his forces were militarily beaten out of Kuwait, he still proclaimed himself as the victor of the 1991 Gulf War, due to his regime surviving in more or less the state it was in before the war, but without control of Kuwaiti. President George H.W. Bush and the U.S. government believed that Hussein would be exiled, overthrown, or killed by his own people, but none of these things happened. President George W. Bush , after 9/11, knew that he wanted Hussein toppled from power by the end of 2001 (pg 52, #14). In the wake of September 11th, 2001, the Bush Administration's top priority was preventing terror organizations, or nations that might harbor them, from gaining access to weapons of mass destruction. This would be the case for generations to come, according to Vice President Cheney. Iraq fit this definition in the mind of President Bush (pg 132, #14).

Because Hussein kicked out weapons inspectors in 1998, the U.S. government could not take the chance of a known mass murderer acquiring weapons of mass destruction or passing them along to terror organizations. In fact, there were those in the Bush Cabinet, including Paul Wolfowitz, Vice President Cheney, Rumsfeld, and Assistant to the

Vice President for National Security I. Lewis (Scooter) Libby, who wanted war with Iraq immediately after September 11th, 2001. Vice President Cheney went to the Middle East to court 10 potential countries for the Iraq invasion, including Egypt, Oman, the United Arab Emirates, Saudi Arabia, Yemen, Bahrain, Qatar, Jordan, Israel, and Turkey. Though these countries did not necessarily support the Iraqi invasion, they still wanted Hussein to be ousted from power in Iraq.

In England, though the U.K. government supported the U.S. decision to go to war with Iraq, over one million English citizens protested against it (pg 108, #18). France, however, voted against the war effort, led by French President Jacques Chirac. In fact, according to Hans Blix in his book *Disarming Iraq*, "As much as the U.S. was bent on invasion, the French, Germans, and Russians were bent on inspections" (pg 206, #17). Blix, a member of Sweden's delegation to the United Nations,. was once Director General of the IAEA (International Atomic Energy Agency) and Chairman of the U.N. Monitoring, Verification, and Inspection Commission for the inspection of Iraq (pg ix, #17).

Because of the disagreement with France, Germany, and Russia on the use of military force to remove Hussein from Iraq, the U.S. and U.K. representatives did not have enough votes within the Security Council for a united force of war against Iraq (pg 218, #17). However, as President Clinton proved during his Administration, it is not necessary to have a united coalition for a military operation, as was the case in Bosnia and Kosovo in the mid-1990s. When ethnic genocide was occurring in these countries, Clinton did not hesitate to send troops to stop the killing, even though this violence did not directly threaten the United States. In fact, President Clinton was hampered by a U.N. vote that prevented him from acting more quickly in Bosnia, which left him unable to prevent the mass slaughter of Bosnian citizens there. To put it another way, a U.N. coalition might cause an unnecessary restraint on the U.S. military to act appropriately when genocide is occurring, making the United States more isolationist than it should be (pg 93, #16).

Bush's philosophy after 9/11 of stopping a Hitler or Stalin before he could cause mass death was called preemption (pg 89, #16). The United States was very concerned about Iraq's ability to strike Israel with Scud missiles. Based on this information, the plan for the invasion of Iraq was a go. C.I.A. Director George Tenet met with Kurdish leaders in Iraq to inform them of the imminent attack on Hussein and his planned expulsion from power (pg 116, #14). By doing so, Tenet was trying to align the U.S. military with the Kurdish forces, and he had tens of millions of dollars to buy friendships (pg 117, #14).

The Kurds, who were definitely persecuted horrifically by Hussein, desperately wanted Hussein out of Iraq (pg 141, #14). General Tommy Franks was the commander of U.S. Central Command, and the four-star

General was given command of the Iraq war plan. General Franks' "hybrid" concept of invading Iraq involved four phases:

1.   Five days to establish the air bridge, which included involuntary enlisting of U.S. commercial aircraft to augment the military airlift to the combat region. Then, eleven days to transport the initial forces to the region.

2.   Sixteen days of air attacks and Special Forces Operations.

3.   One hundred twenty-five days of decisive combat operations. At the beginning of the 125 days, they would try to get a division inside Iraq, and within a week another division of ground forces.

4.   Stability operations of an unknown duration (pg 146, #14).

General Franks estimated to President Bush that it would take 265,000 troops to govern Iraq once Hussein was ousted from power. He said that the number of troops could be lowered over time. Military targets in Iraq included Ba'ath Party headquarters in Baghdad. Secretary of State Colin Powell, however, was concerned about the American military plan for Iraq. Powell, in a private meeting with President Bush, warned him of the dangers of occupying Iraq. Powell told Bush that he would become responsible for 25 million Iraqis and all of their hopes and problems. Powell also warned the president of the threat of destabilization of friendly regimes, places such as Saudi Arabia, Egypt, and Jordan. Secretary of State Powell said to President Bush, "This will become the first term" (pg 150, #14).

Powell continued his warnings to President Bush by pointing out that Iraq had never been a democracy, and that most of the U.S. Army deployed worldwide would have to be stationed in Iraq, instead of being more spread out (pg 151, #14). Finally, contrary to the go-it-alone policy, Powell strongly suggested to Bush that a U.N. mandate would strengthen the American position and win new American allies.

However, President Bush countered Powell's proposals by insisting on protecting Americans first, above all else. Vice President Cheney disagreed with Powell's U.N. suggestion, as did Defense Secretary Rumsfeld. Some former Administration officials, such as Secretary of State James Baker, Brent Scowcroft, and Henry Kissinger, urged the president to gain foreign support before any military action was implemented (pg 163, #14). Yet Vice President Cheney would stand firm on the political platform that Iraq had nuclear weapons and was ready to attack the United States and its allies. The longer America waited, Cheney said, the more America would be at risk (pg 165, #14).

Another Al-Qaeda terrorist attack occurred in a Bali nightclub on October 12th, 2002, mercilessly killing 202 innocent people who were mostly Australians. This was another example of the elusiveness of the evil terror organizations that democracies were fighting worldwide. Accordingly, U.N. Security Council Resolution 1441 was adopted on November 9th, 2002, which required Iraq to cooperate with U.N.

inspectors—"immediately, unconditionally, and actively"—in their search for forbidden weapons.

The reason for the passage of Resolution 1441 was for the United Nations to give Iraq one last chance to cooperate with the U.N. inspectors (pg 93, #17). This was an opportunity for Iraq to comply with inspectors so that the war could be avoided, and the eventual lifting of economic sanctions against Iraq could have, at least, been remotely possible. Blix pointed out that South Africa was forced to go through similar U.N. inspections, with which they complied. Despite this, Bush was very skeptical of Hans Blix, and felt that Iraq was a threat that had to be dealt with. Accordingly, on October 10th, 2002, the U.S. House of Representatives voted to authorize President Bush to go to war in Iraq by a vote of 296 to 133 (pg 130, #19).

Though Powell had advised against it, President Bush told the Secretary of State on January 13th, 2003 that he was taking the U.S. into war with Iraq. Though Bush had said publicly that war was his last choice for Iraq, it was really his first (pg 241, #14). However, Bush ran into considerable opposition from the other permanent U.N. Security Council members, such as France, Russia, and Germany, which were against the military invasion of Iraq. The three leaders Chirac, Vladimir Putin, and Gerhard Schroeder issued a statement opposing the war and expressing the desire of disarming Iraq peacefully. These countries' disagreement with the war was fueled by Hans Blix's presentation to the U.N. Security Council that his inspectors had inspected 300 sites in Iraq, including industrial sites, ammunition depots, research centers, universities, Presidential sites, private houses, missile production facilities, military camps, and agricultural sites, and had found no prohibited weapons (pg 317, #14).

Even Mexico's President Vicente Fox and Chilean President Ricardo Lagos did not back Bush with a U.N. vote for war (pg 345, #14). However, by then, despite President Bush saying on a primetime news conference that he had not yet made up his mind about military action in Iraq, Cheney, Rice, Powell, and Rumsfeld knew that he had. On December 19th, 2002, Bush declared the country of Iraq to be in "material breach" of Resolution 1441.

By the end of February 2003, there were 200,000 U.S. troops stationed in the Gulf region, ready for war (pg 146, #17). The inspection process in Iraq lasted only three months (pg 268, #17). Bush set the deadline for Hussein to vacate power for March 17th, 2003, or the consequences would be war. England's Prime Minister Tony Blair successfully won a mandate for the Iraq War by a vote of 396 to 217, which included England's Tory support. When Hussein and his sons failed to withdraw into exile from Iraq, and the deadline had come and gone, the war against the Hussein regime began on March 19th, 2003. President Bush said in his war declaration, "For the peace of the world, and the benefit and freedom of the Iraqi people, I hereby give the order to execute Operation Iraqi Freedom. May God bless the troops" (pg 379, #14). The fedayeen is the name of the enemy forces who fought American and British forces in Iraq (pg 232, #19).

Turkey had rejected America's request to attack Iraq through Turkish territory, as America had previously done in the first Gulf War in 1991. The first place for America to attack was the Rumaila oil fields, which were parallel to the Iranian border and approximately 50 miles below Basra. Over 300 oil wells pumped over two million barrels of oil a day (pg 146, #15). The siege of the Rumaila oil fields was successful.

The next military target was Basra. The city is approximately two kilometers wide, with some additional suburbs to the south (pg 176, #15). The U.K. military, besides its excellent naval forces, focused on counter-insurgency tactics, which dates back to its experience in the battles in Northern Ireland, in which it fought from street to street. The siege of the city began on March 23rd, 2003, by British Special Forces units, like the Royal Marine Commandos Special Boat Service teams. British troops excel in targeting terrorists by acting in small groups, without causing harm to the civilian population.

By April 8th, 2003, the city of Basra was successfully taken by the British. The 82nd and 101st Airborne Divisions for the U.S. military also did well in securing the city of Najaf (pg 157, #15). However, the Battle of Nasiriyah was bloody, and many U.S. Marines were killed or wounded (pg 254, #19). Next was Baghdad and the area called Karbala Gap, which ran near Highway 8. This road was near Baghdad (Saddam) International Airport. The Iraqi war material was no match for the vastly superior American arms and weapons (pg 193, #15). The U.S. military had Abrams tanks and Bradley fighting vehicles, which were vastly superior to the Iraqi weapons (pg 195, #15). Yet Baghdad was full of fanatics, including the Republican Guard, the fedayeen, the regular army, and foreigners from outside Iraq (pg 193, #15). The fedayeen were Saddam's military men, many of whom were converted into the insurgency (pg 505, #19). Some Syrians fought with the Iraqi soldiers (pg 197, #15).

Hundreds of Iraqi soldiers were killed in the battle for Baghdad. The statue of Hussein fell in Baghdad on April 9th, 2003 (pg 202, #15). On May 1st, 2003, President Bush landed onboard the *U.S.S. Lincoln*, stationed in San Diego, California, to declare major combat missions in Iraq as being officially over (pg 412, #14). Iraq was then transformed from a dictatorship into a democracy. Hussein, who thought that he was the modern-day Nebuchadnezzar, was eventually caught in hiding by the United States outside Tikrit on December 13th, 2003 (pg 425, #14).

The reason for war with Iraq from the U.S. standpoint seemed to shift from eliminating Iraq's weapons of mass destruction, to the removal of Saddam Hussein from power, to creating an Iraqi democracy. In fact, C.I.A. Director George Tenet once declared the case for weapons of mass destruction existing in Iraq as a "slam dunk," as did multiple others in the Bush Administration. Yet no one was fired for this mistake, with President Bush himself saying that it is better to be safe than sorry. Therefore, President Bush offered no apologies for declaring war in Iraq.

As a result of the second Iraq War success for America, Libyan leader Muammar Qaddafi agreed to give up all weapon-of-mass-destruction programs in his country in December 2003. Yet according to Hans Blix, the U.S. and British governments greatly exaggerated the threat of war in order to declare it, thus causing these two countries to lose credibility on the international stage (pg 271, #17). Blix also blames the United States and the United Kingdom for diminishing the authority of the United Nations (pg 274, #17). However, it is agreed internationally and domestically that the ousting of Saddam Hussein from Iraq has made the world a safer place and rid the world of one of its most wicked dictators in modern times (pg 218, #15).

Iraq could be described in time as the first Arab democracy. Ayad Allawi, a Shi'ite, was elected Iraq's first Prime Minister in 2004 (pg 95, #23). Also, imposed democracies, such as the one in Iraq, have worked in the past with countries such as Japan, Italy, Germany, Austria, Grenada, Panama, and the Dominican Republic (pg 98, #16). It is true that exporting democracy to other countries has furthered America's vital interest, because democracies have never declared war against each other (pg 105, #16). Therefore, promoting democracy worldwide does make the world more congenial to countries like America and England (pg 106, #16). The Prime Minster of Iraq from May 20th, 2006 to 2014 was Nouri al-Maliki, who took over for Ibrahim al-Jaafari, who replaced Allawi.

However, it is misleading to suggest that the war against terror in Iraq and Afghanistan is anywhere near over. In 2004, President Bush, General George Casey, and Secretary of Defense Rumsfeld had to extend deployments of U.S. troops because of the deteriorating security situation in Iraq (pg 65, #23). There had been an average of 150 terror attacks a day in Iraq. Also, in 2004, terrorists in Iraq had 4 safe havens: Fallujah, Najaf, Samarra, and Sadr City (pg 17, #23). General David Petraeus took over for General George Casey as commanding general in the Iraqi theatre on February 10th, 2007. The new Shia government was openly discriminating against the Sunni population once it took power in Iraq. As the Sunnis did not participate in the January 30th, 2005 election in Iraq, many formed the heart of the insurgency (pg 383, #22). Sunni insurgent attacks on coalition forces and Iraqi civilians steadily rose each month since May 2003 (pg 472, #22). U.S. and coalition forces are continuing to die in the theatre of combat in these war regions. The U.S. military is still fighting insurgents coming in from Syria and Iran, and soldiers are continuing to be killed from roadside explosions, suicide bombings, and mortar attacks. As of June 29, 2016, there have been over 4,424 U.S. soldier combat fatalities in Iraq, with approximately 31,000 wounded. Iraqi fatalities have been reported to be 100,000 since the war began. Iraq's landmark constitution was passed by the majority of Iraqi voters by a margin of 78% to 22%. The Sunni Arab population, of which Hussein was a member, failed to gain enough support to defeat the referendum, while the Kurds and Shi'ites strongly supported it.

However, U.S. soldier deaths continue to climb, and the insurgency still continues to fight the democratically installed Iraq government, with the exception of the Al-Qaeda leader in Iraq, Al-Zarqawi. He was responsible for beheading hostages and murdering troops stationed in Iraq, and he was killed by coalition forces, who bombed his hiding area in June 2006. The United States still has to straighten out wrongful conduct at the Abu Ghraib and Guantanamo Bay prisons committed by U.S. soldiers against prisoners of war, defying the Geneva Conventions that we have honored ever since their inception. Our U.S. soldiers will be, and have been, disciplined for these events. Insurgents are still killing Iraqi civilians and American/U.N. coalition soldiers in roadside bombings, suicide car bombs, and guerilla-style ambushes almost daily. The country of Iraq is not yet politically and internally strong enough to control the insurgents, who wish to destroy democracy and all that it stands for. The U.S. military has been criticized for not properly planning to counter the insurgents and conduct anti-terrorist operations as soon as Baghdad fell (pg 501, #19). Retired General Eric Shinseki said in 2003 that there were not enough U.S. troops in Iraq. President Bush said on August 22nd, 2007, that, "unlike in Vietnam, if we withdrew before the job is done, this enemy will follow us home" (pg 378, #23). He also added that withdrawal from Iraq would leave "a region already known for instability and violence under a shadow of a nuclear holocaust" (pg 379, #23). Some experts, contrary to President Bush's vision of Arab democracy, see this as a flawed foreign policy initiative. Defense Intelligence Agency expert Derek Harvey, a strategic adviser who reported directly to General David Petraeus, said that "even if Iraq turned out well in the end ... it would not rescue the Bush legacy. For too many years, from 2003 to the end of 2006, the president had not been frank about the costs, duration, and challenges of what had been undertaken in the Iraq War" (pg 428, #23). Bush supported a surge of an additional 30,000 to 40,000 troops to reduce the violence and chaos in Baghdad and the surrounding areas, and it has largely been effective. Violence in Iraq did drop dramatically because of the rise in U.S. troop levels in Baghdad, coupled with General Petraeus' security plan. Also, 4,000 additional U.S. Marines were sent to secure Anbar Province in western Iraq, and U.S. intelligence agencies shared top-secret information that was helpful in reducing violence (pg 380, #23). However, the cost of funding the war in Iraq was approximately $2 billion a week (pg 222, #23). There are still 5,500 U.S. troops in Iraq and 11,000 in Afghanistan at present.

Operation Neptune Spear
Osama bin Laden, the mastermind and chief financier of the September 11th, 2001 attacks on the United States, was killed by U.S. Navy Seal Team 6 on May 2, 2011. Twenty-three Seals in two Black Hawk helicopters raided the compound that was located in Abbottabad, which is north of Pakistan's capital of Islamabad. Pakistan's time zone is 9 hours ahead of Washington, D.C. The raid lasted for approximately 40 minutes. Five people, including Osama bin Laden, were killed in the raid, including Osama bin Laden's son. One of the helicopters involved in the raid crash landed in the compound, but no one was hurt and all Seals successfully escaped in the one working helicopter after the mission was successfully carried out. Robert O'Neil claims to be the soldier who personally killed Osama bin Laden.

Osama bin Laden's body was also taken out of the compound by the Seals into Afghanistan for identification, then buried at sea less than 24 hours after his death, which is in accordance with Islamic traditions. Also, no country would accept the terror leader's remains. When American military intelligence reviewed Osama bin Laden's computer files collected during the raid, it was discovered that he was planning more attacks in the United States, including on the anniversary of 9/11. He also was plotting to have President Barack Obama assassinated. Among items confiscated in Osama bin Laden's compound by the American military were 5 computers, 10 hard drives, and more than 100 storage devices, including discs, DVDs, and thumb drives. The break in finding Osama bin Laden came when the C.I.A. tracked the terror leader's courier to the Abbottabad compound, which was protected behind large security walls in a populated residential neighborhood. The compound was less than a mile from an elite Pakistani military academy in August 2010.

At 11:30 p.m. on May 2, 2011, President Obama went on national television to announce the successful raid on the compound that resulted in the death of Osama bin Laden. The president said, "Justice has been done." Crowds gathered at spots all over the United States, including the White House, Times Square in New York City, and Ground Zero, to celebrate (History.com, Wikipedia, CNN library online).

Iraq
The Islamic State in Iraq and Al-Sham (I.S.I.S.) seized control of the Iraqi city of Mosul in June 2014, and claims itself to be a Sunni caliphate. Al-Qaeda recruit Abu Musab al-Zarqawi is credited with beginning the ISIS movement in Iraq in 2003. Egyptian doctor Ayman al-Zawahiri was in Al-Qaeda with Osama bin Laden. Al-Zarqawi recruited some of Saddam Hussein's men to join him. The counter-insurgency in Iraq grew strong in 2004. Al-Zarqawi and Al-Qaeda disagreed in tactics and frequently argued with one another. Al-Zarqawi was killed in a U.S. air strike in June

2006, but his followers declared themselves to constitute the Islamic State of Iraq without consulting AI-Qaeda.

Former Iraqi Prime Minister Nouri al-Maliki was a Shi'ite Muslim and displaced Sunnis from power in Iraq. Bashar al-Assad in Syria allowed I.S.I.S. to grow in his country to show the West why he should stay in power. This author recommends a resolution by the United States, Russia, Saudi Arabia, Turkey, and Iran to cooperate in a peaceful post-Assad government once he leaves power, or I.S.I.S. will grow stronger. Al-Maliki hated Saddam Hussein's regime of Ba'athist Sunnis; I.S.I.S. gained strength in Raqqa, Syria and openly recruited Sunnis there. Al-Maliki forced attacks on Sunni protesters in Iraq, killing hundreds, and I.S.I.S. then gained a stronghold there and ran freely between Syria and Iraq.

Al-Maliki hit the Sunnis hard again in December 2013. This provoked a huge Sunni counterattack against Al-Maliki. Then, I.S.I.S. defeated the Iraqi Army in Ramadi, Fallujah, Mosul, and Tikrit, the hometown of Saddam Hussein, and has employed Ba'athist Party military leaders that Al-Qaeda never had. President Obama acknowledged that the United States underestimated the threat posed by I.S.I.S. and overestimated Iraq's ability to defeat it independently. A counter-insurgency of Kurdish forces have been attacking I.S.I.S. strongholds in Fallujah and Mosul, as well as Raqqa; I.S.I.S. claims to represent the Sunnis. As of October 28, 2016, U.S. Forces lost 2,386 soldiers in the 15-year campaign, with 20,000 wounded in Afghanistan. The current leaders in both Iraq and Afghanistan, respectively, are President Ashraf Ghani and Prime Minister Haider al-Abadi, both of whom were elected to power in 2014.

Bibliography

1. *From Beirut to Jerusalem*, by Thomas Friedman (previously listed).
2. *Against All Enemies: Inside America's War on Terror*, by Richard Clarke (Simon & Schuster, 2004), disc 6.
3. *The Complete Book of U.S. Presidents*, by William A. Degregario (previously listed).
4. *My Life Volume 2: The Presidential Years*, by President Bill Clinton (Vintage Books, 2004-2005), disc 10.
5. *The 9/11 Commission Report: Final Report of the National Commission on Terrorist Attacks Upon the United States*, Commission members: Thomas H. Kean, Chair; Lee H. Hamilton, Vice-Chair; Richard Ben Veniste; Bob Kerrey; Fred F. Fielding; John F. Lehman; Jamie S. Gorelick; Timothy J. Romer; Slade Gorton; James R. Thompson (W.W. Norton, 2005), disc 7.
6. *The Cell: Inside the 9/11 Plot and Why the F.B.I. and C.I.A. Failed to Stop It*, by John Miller and Michael Stone, with Chris Mitchell (Hyperion Books, 2002), disc 5.
7. *Bush at War*, by Bob Woodward (Simon & Schuster, 2002), disc 4.

8.*Understanding September 11th: Answering Questions About the Attacks on America*, by Mitch Frank (Viking 2002), disc 4.

9. *Hunt for bin Laden*, by Robin Moore (Random House, 2003), disc 8.

10."Tora Bora Falls, but No bin Laden" (*Christian Science Monitor*, December 17th, 2001).

11. National Geographic video: "Afghanistan: Untold Story of a Land and Its People," disc 3.

12. *Al-Qaeda's Great Escape: The Military and the Media on Terror's Trail*, by Phillip Smucker (Brassey's Inc, 2004), disc 10.

13. *The Price of Loyalty*, by Ron Suskind (Pulitzer Prize winner) (Simon & Schuster, 2004), disc 8.

14. *Plan of Attack*, by Bob Woodward (Simon & Schuster, 2004), disc 6.

15. *The Iraq War*, by John Keegan (Knopf, 2004), disc 8.

16. *The War Over Iraq: Saddam's Tyranny and America's Mission*, by Lawrence F. Kaplan and William Kristol (Encounter Books, 2003), disc 8.

17. *Disarming Iraq*, by Hans Blix (Pantheon Books, 2004), disc 8.

18. *War on Iraq: What the Bush Team Does Not Want You to Know*, by William Rivers Pitt, with Scott Ritter (Context Books, 2002), disc 8.

19. *Cobra 2: The Inside Story of the Invasion and Occupation of Iraq*, by Michael R. Gordon and General Bernard E. Trainor (Pantheon Books, 2006), disc 10.

20. *Understanding Iraq*, by William R. Polk (HarperCollins, 2005).

21. *Bosnia: A Short History*, by Noel Malcolm (New York University Press, 1994).

22. *State of Denial: Bush at War Part 3*, by Bob Woodward (Simon & Schuster, 2006).

23. *The War Within: A Secret White House History, 2006-2008*, by Bob Woodward (Simon & Schuster, 2008).

Chapter 15
Conclusion

To summarize, what do all of these chapters about past war conflicts tell us about our country, the United States? What do they tell us about our democracy and the separation of powers that our Constitutional forefathers foresaw when they formed this unique experiment of multiple points of view united under a federal beacon of freedom? Will this nation return to its full dominance on the world stage as a global power, yet remain the example of diversity and confidence that we took for granted before the September 11th, 2001 attacks on our nation? Will terrorism ever be truly defeated? Where do we go from here?

The first expression that comes to mind is that freedom came at a very high price throughout American history, especially when the odds were heavily stacked against America's quest for the right to live in freedom. Our own Constitution has evolved during American history, to include all of our citizens participating in the election of our leaders of government, when initially only white male land owners could obtain such liberties. In fact, African-Americans could not vote for their representatives until the passage of the Fifteenth Amendment in 1870, soon after America's brutal civil war, which claimed the lives of 620,000 Americans. Women couldn't vote for their representatives until the passage of the Nineteenth Amendment in 1920. Even Native Americans couldn't exercise their right to vote until 1924, and Washington, D.C. residents couldn't vote until the passage of the Twenty-third Amendment in 1961.

It can be said that our own Constitution and the laws that govern our land are flexible enough to evolve over time, with technology, increasing populations, and new, rapidly changing threats posed against our citizens and law enforcement. This must be done in order to best protect and safeguard our well-being. A great portion of the wars that were fought in American history reflect the fruits of victory from military campaigns that were successful in achieving freedom and liberty at home and in liberating other nations around the world. Some of these nations that were once our bitterest of adversaries have become our closest of allies, both politically and militarily (for example, England, Japan, and Germany). Despite this, as the world witnessed on September 11th, 2001, terrorists and enemies of freedom can successfully mutate and infiltrate our free society to inflict

251

their radical ideology of destruction and terror, unless law enforcement, the U.S. military, and intelligence agencies can stop them before they strike.

In the case of 9/11, the dots were not connected to have prevented this nefarious incident, and over 3,000 innocent people of all races and religious denominations were murdered as a result. Since 9/11, other acts of terror have occurred in Bali, Spain, England, San Bernardino, and Paris. Our evolving terrorist enemy is resilient and is always on the offensive in other parts of the world in its effort to destroy freedom. The terror organizations that America and our allies fight today are described as "sophisticated, patient, and disciplined."

I wrote this book to honor the memory of the people we lost on that tragic day of September 11th, 2001. That day represented the largest loss of civilian life due to terror in U.S. history, and a day that, as President Franklin Roosevelt said, was a "day that will live in infamy." Yet, as we learned on that horrible day, in this new era of terror attacks in the post 9/11 world, we are all primary targets to our enemies. While the U.S. military presence in both Iraq and Afghanistan has diminished in 2017, new problems have arisen in places in the Middle East that have created a political and terror quagmire in the region, namely in Syria and parts of Iraq that have fallen under the Islamic State terror group known as I.S.I.S. (Islamic State of Iraq and Al-Sham).

The Syrian civil war, which has been going on since 2011, is the deadliest conflict the 21st century has witnessed thus far. The Syrian death toll from this conflict is estimated to be more than 465,000, with another 500,000 injured and 12 million Syrians displaced from their homes. When the war started in 2011, the Arab Spring had toppled Tunisian President Zine El-Abidine Ben-Ali and Egyptian President Hosni Mubarak. By March 2011, protests began by a group of teenagers, who were summarily rounded up, tortured, and killed by the regime of Syrian President Bashar al-Assad. Assad would go on to kill hundreds more protesters, while imprisoning hundreds of others. By July 2011, military defectors from Assad formed the Free Syrian Army that attempted to overthrow the Assad government, leading to a civil war. Previously, Bashar's father Hafez al-Assad ordered a military crackdown in 1982 against the Muslim Brotherhood, killing an estimated 10,000 to 40,000 people.

Most Syrians are Sunni Muslim, but the country's security establishment has long been dominated by members of the Shi'ite Alawite sect, of which Assad himself is a member. A severe drought plagued Syria from 2007 to 2010, making 1.5 million of its inhabitants migrate from the countryside to the cities, which further exacerbated social unrest and extreme poverty. The Obama Administration hesitated to involve itself deeply into the Syrian conflict, even when Assad used chemical weapons against his own people in 2013. However, the Obama Administration had stated repeatedly its strong opposition to the Assad government. Obama said that the chemical attack that the Assad regime carried out was the "red line" that would prompt a possible U.S. ground invasion. Although President Obama didn't order an invasion of Syria, he did order the U.S. military to bomb targets that were known to the intelligence community as I.S.I.S. strongholds.

Since 2014, Russia, led by President Vladimir Putin, has backed the Assad regime in Syria, including voting against 8 Western-backed resolutions of the U.N. Security Council. Russia is responsible for bombing rebel group strongholds that are attempting to topple Assad. Russia has also deployed military advisers to shore up Assad's defenses. Several Arab states have provided weapons and material to the rebel groups in Syria. The Assad regime is backed by the Shia majority from Iraq and by Lebanon-based Hezbollah. Sunni majority states, including Turkey, Saudi Arabia, and Qatar, strongly support the Syrian rebels.

I.S.I.S. forces captured portions of Iraq and Syria in 2013. They used social media to carry out brutal executions, including beheadings on camera. Kurdish forces are fighting Assad for autonomy in northern Syria, just as they are in Iraq. Assad's forces did successfully recapture the Syrian city of Aleppo in 2014. With much of Syria in ruins, millions of Syrians have fled to surrounding countries, including Lebanon, Turkey, and Jordan. Rebuilding Syria after the war ends will be a lengthy, difficult process. After all, Assad has been in power in Syria since 1994. Assad's strength largely comes from Russian, Chinese, and Iranian support.

Sizable portions of secular Arab Syrians and religious minorities of Alawites, Assyrian Christians, and Druze people are firmly in Assad's camp. Three groups that are ferociously fighting to remove Assad from power in Syria are the Free Syrian Army, made up mostly of defectors from Assad's forces; the Y.P.G., a militia group called the People's Protection Units; and the Syrian Democratic Forces (S.D.F.) These three groups became a strong defense against I.S.I.S. combatants. However, I.S.I.S. emerged as a key political and military force in Syria in 2014. Unlike other insurgent forces, it didn't fight Assad or his regime. Instead, it opportunistically claimed large swaths of uncontrolled land and declared an independent caliphate state, becoming the chief source of radicalization that threatens Western society, even passing Al-Qaeda.

Iran has been a long-standing ally of the Assad regime because of its sectarian, political, and economic interests. The Alawites, who are associated with the Shi'ite Muslims, include Assad and his entourage as members. Iran's government is Shia dominated as well. Because of this, Syria is an important corridor for Iran to press its influence over Lebanon's Shi'ite militia Hezbollah and to provide access to the Mediterranean. Iran's regional ambitions require the continuation of the Assad regime. Worried about Iran's growing influence in the region, Saudi Arabia and Qatar have supported Salafist insurgent groups. This is because they fear the spread of I.S.I.S. ideology and popularity within their geographical realm, and the Saudi government has supported U.S.-led air strikes on I.S.I.S. since 2014. The country of Saudi Arabia, led by King Salman bin Abdulaziz Al Saud, is composed mostly of Sunni Muslim residents.

In the *New York Times* article "In the Rise of ISIS, No Single Missed Key, but Many Strands of Blame," by Ian Fisher (#16), it states that by the time of the U.S. military withdrawal from Iraq in 2011, I.S.I.S. was thought to be on its way to defeat. This was because of the successful U.S. troop surge and the Sunni insurgency. Al-Qaeda's leader in Iraq, Abu Musab al-Zarqawi, was killed by U.S. forces in 2006. After his death, Abu Bakr al-Baghdadi became the new leader of I.S.I.S. In fact, I.S.I.S. was formed 4 months after Zarqawi's death. When U.S. troops left Iraq in 2011, they felt that Al-Qaeda was beaten in Iraq. By the end of 2014, 4,497 U.S. troops had been killed throughout the duration of the war there, which began in 2003. But when Prime Minister Nouri al-Maliki, who was a Shi'ite Muslim, took over a mostly Sunni population, there was tension and problems.

Meanwhile, President Barack Obama's second term was well under way after his successful re-election in 2012. Although he was unsuccessful in getting his gun legislation or his immigration reform proposals passed, he was successful in getting Iran to sign a deal to get its government to stop developing nuclear weapons in exchange for some sanctions being lifted against Iran. Obama was also successful in normalizing and re-opening diplomatic ties with Cuba. In his first term, along with the successful killing of Osama bin Laden by the Navy Seals, Obama ended the 2008 recession, won the 2009 Nobel Peace Prize, reformed health care, and regulated big banks. Obama lowered unemployment from almost 10% when he took office to 4.7% by the time he left office in January 2017. Despite this, the despicable amounts of gun violence continued during Obama's presidency, including the Orlando nightclub tragedy; multiple school shootings, including the massacre at Sandy Hook elementary school in Newtown, Connecticut; the San Bernardino, California shootings; and the South Carolina church massacre. President Obama was criticized for underestimating the rise of I.S.I.S. and overestimating the ability of the Iraqi military to fend off I.S.I.S. President Obama said that a lasting peace can only be secured with a political solution. Obama said, "What we have to do is we have to come up with political solutions not only in Iraq and Syria, but in the Middle East generally, that arrive in the combination between Sunni and Shia populations that, right now, are the biggest cause of conflict not just in the Middle East, but around the world."

Article 2, section 2 of the U.S. Constitution declares that the President of the United States is the Commander-in-Chief of the armies and the navies. However, according to Article 1, section 8, it is the Congress of the

United States that has the vested powers to declare war against other nations or entities. Thus, in order for a president to officially declare and implement a war, he or she must get the measure passed through the House of Representatives. In 1973, Congress passed the War Powers Resolution, which required the Chief Executive to consult with Congress before committing troops in any hostilities. However, this act was sometimes ignored by presidents after 1973. Importantly, Article 4, section 4 says that the United States guarantees its citizens a republican form of government and protection from foreign attackers, including domestic violence. As President Lincoln demonstrated during the horrific situation of the Civil War, in Article 1, section 9 of the Constitution, the president does have the power to suspend habeas corpus in cases of rebellion or invasion if the public safety requires such action, which during the Civil War it most certainly did.

It is the presidential oath, after all, that states that the president must do everything possible to "preserve, protect, and defend the Constitution of the United States." Since its inception in 1776, the United States has become the world's largest economy and superpower. America's enormous wealth and resources have catapulted us to our current number one status when it comes to responding to suffering and disaster anywhere in the world, and this practice must continue to thrive in our nation's future. According to Dr. Henry Kissinger in his book, *Does America Really Need a Foreign Policy?* he states that the United States is "enjoying a preeminence in the global society that is assuredly unrivaled by all of the former greatest empires in history, and it is truly unparalleled" (#3). As we have seen from the wars in the previous chapters in this book, it is clear that the American military has paid the enormous cost of our freedom through bravery and the intense warfare it endured on our behalf. These sacrifices that our soldiers, past and present, have given is something that we all must remember in our daily lives, and we, as citizens, must never forget them.

America has adopted the Wilsonian philosophy that states that the defense of the principle of human rights and freedom globally, even by force, was to become the general practice of the American national interest. In Newt Gingrich's book, *Winning the Future*, he tells us that Islamic terrorists and rogue dictatorships acquiring weapons of mass destruction will be our national security's biggest threat for the foreseeable future, and that the United States must commit to a war to defeat these terrorist groups. Gingrich says, "A major part of our homeland defense build-up must be a large boost in the number of F.B.I. and border patrol agents, as well as enlargement of the Coast Guard" (#4).

As far as our new policy goals to keep our security protected from terrorism, our post-9/11 goals for the 21st century should be as follows:
1. To preserve our national security
2. To protect individuals from harm

3. To preserve our national way of life
4. To ensure the viability and prosperity of the economy
5. To bring justice and peace to various parts of the world
6. To heal the pains of national tragedy and personal loss

We can thank our past presidents, who created the protection that we depend on today with agencies such as the North American Treaty Organization, the United Nations, and the Central Intelligence Agency. All 3 of these agencies were created by President Harry Truman. Also, President John Kennedy's creation of the Peace Corps in March 1961 set out to improve conditions in underdeveloped countries by enlisting volunteer members to teach and provide assistance to less fortunate people. On the other side of the aisle, President Richard Nixon not only initiated the nuclear reduction treaties with the Soviet Union, but also helped to bring about the destruction of stockpiles of weapons and a ban on any reproduction of such items.

Because of the tragic events of 9/11 and other acts of terror that are committed against innocent civilians globally, our world seems to have changed into a domain of fear and mistrust. People now have extreme anxiety about our safety, and wonder who next will suffer these atrocities implemented by those who hate our way of life. Terror cells will go to every effort to destroy and end as many lives as possible to promote their evil agenda. These terrorists despise freedom and wish for their distorted views of Islam to become the new global dominance on the world stage. On November 13, 2015, 3 I.S.I.S. members killed 137 innocent people in Paris, France, including many at the Bataclan theatre who were attending a rock concert by the American group Eagles of Death Metal. This is not the first, and probably not the last, war over ideas and misunderstanding of various religions and their scriptures.

According to the book *Abraham: A Journey to the Heart of Three Faiths*, by Bruce Feiler, he asks the question, "So what is the message of Genesis after September 11, 2001? Why do religious people act the way that they do? It is because of a lack of modesty" (#5). To understand the role that religion can play in both politics and culture, especially in foreign governments, I believe that it is very important to examine what various religious doctrines say about their individual philosophies and divinely inspired beliefs.

According to the book, *God*, edited by Jacob Neuser, he states that "each religion forms a system with its own definitive traits" (#6). On a global scale, there are five major religions that encompass a large amount of the world's population: Christianity, Judaism, Islam, Buddhism, and Hinduism. In fact, the three religions of Christianity, Judaism, and Islam have a shared ancestor in Abraham. Accordingly, Abraham is the father of 12 million Jews, 2 billion Christians, and 1 billion Muslims around the world. Despite Abraham's massive influence on these three major world religions, the only historical record of him exists in the *Bible* and *Koran*.

According to the story in Genesis, Abraham had children later in his life. While Abraham had a son, Isaac, from the union with his wife Sarah, he first had a son from a union he had out of wedlock with Hagar, and his name was Ismael. Abraham established the 12 tribes of Israel with his son Isaac. However, Isaac's son Jacob is credited with the establishment of the Islamic faith. The *Koran* states that Ismael relocated to the area of Canaan, which is located in Syria. A very important pillar of Islam for all Muslims is the Hajj, which means that all Muslims must make a visit to the holy city of Mecca at least once in their lifetimes. The character of Abraham, ironically enough, is just as important to Muslims as he is to Jews and Christians. In Feiler's book on Abraham, he says that "Abraham is the father of Jews, Muslims, and Gentiles alike. Any person who shows faith is a descendent of Abraham."

According to the *Koran*, "God is lenient, merciful." The holy *Koran* also states that God guideth not the evil-doers, God loveth no infidel or evil person." The word God in Arabic is Allah, the same God that Jews and Christians believe in. It is said that the faiths of Islam, Christianity, and Judaism are similar to each other in the deep belief in one absolute God and the pure faith in divine existence. All three of these religions teach that it was Abraham who invented monotheism. This philosophy differs from that of the Hindu faith, which teaches a concept called polytheism, or the belief in one or more gods. In Hinduism, there exists the idea of *samsara*, which is the endless cycle of death and re-birth. Members of the Buddhist faith also agree with the Hindus in their belief in reincarnation. Yet the difference between Buddhism and Hinduism is that Buddhists place more emphasis on one's personal actions then in divine guidance and/or intervention. Therefore, morality is something that is considered very important to Buddhists.

Christianity expresses the love of God through the chosen messiah, who is Jesus. Christianity focuses on the union of the Trinity (Father, Son, Holy Ghost). However, the philosophy still emphasizes the words and teachings of Jesus. Therefore, in order to complete God's teachings and receive His love, it is important to imitate Christ. Yet this practice of Jesus' teachings was misinterpreted many times in the course of history. For example, the Crusaders' actions of persecution against non-Christians, and King Ferdinand and Queen Isabella's persecution of all non-Catholics in Spain in the year 1492, did directly conflict with the message of Jesus. As a result, all non-Catholics were either baptized or tortured/killed during this period of the Inquisition. Christians were massacring people who didn't share their beliefs, which is a direct contradiction of the message of Christianity. Christianity is divided among three unique practices: Eastern Orthodox, Roman Catholic, and Protestant. Christianity still remains an incredibly popular and influential religion globally, with an estimated one third of the world's inhabitants calling themselves members. While Christianity is spread throughout the many regions of the world, large numbers of the religion live in Africa, South America, and Asia.

According to the U.S. Constitution, all of the religions that I have listed, including more obscure, diverse, and ethnic examples (such as Native American religious practices), are legally protected by the Constitution. According to the First Amendment of this important document, it states that "Congress can make no law prohibiting the freedom of speech, religion, the press, or the right to peacefully assemble." While the Articles of Confederation covers the framework of legislative responsibilities of the separation of powers (executive, legislative, and judicial), the Bill of Rights addresses the aspects of individual rights. These include the aforementioned First Amendment, the Second Amendment, which is the right to bear arms, the Fourth Amendment, which is the right not to be unreasonably searched without a warrant, and the Fifth Amendment, which states that no person shall be prosecuted without due process of law. The Sixth Amendment provides the right to have a fair and speedy trial, and the Ninth Amendment guarantees that American Constitutional rights are for all of its citizens. Because of all the additional amendments that have been passed by Congress, it is clear that our own Constitution is a document that needed to and has evolved over time. This is precisely why the framers constructed the Constitutional document in the manner that they did, in order to evolve it and maintain its strength to meet any social or political changes that may occur over time.

It does seem probable for a country that is new to the democratic process of government that it will also encounter similar changes and pass new laws that are necessary, just as we did. These laws will change over time to reflect the will of a nation's citizens and to better represent the will of its majority. This includes differing political parties, religions, financial backgrounds, employment, military service, relations with foreign leaders, etc. As Dr. Robert Remini states in his book *The House: The History of the U.S. House of Representatives*, "throughout our nation's history, change has been a constant" (#8). The United States became a model for a successful democratic government that many nations across the globe have transformed into their own thriving forms of democracy. The U.S. Constitution guarantees that all of our citizens can enjoy the right to life, liberty, and happiness.

However, when times of war arise for our nation, responsibility lies with the executive branch, in cooperation with Congress, which must carry the heavy burden of sending our troops into harm's way to meet the victory objectives for each individual campaign. Each president that I have researched for this book had unique and daunting circumstances that he had to face in order to achieve both military and political victories in war. Each executive used his power to the fullest extent possible to achieve victory, whether or not he was 100% successful. Below, I briefly summarize each wartime president's decisions and courses of action, including impending consequences of each one's particular wars on American history, in chronological order:

1.   The Revolutionary War was a fight involving the American colonists, who protested unfair taxation and oppression from England. Although the war campaign lasted from 1775 until 1783, the outstanding accomplishment achieved by the Colonial Army against a vastly superior fighting force, England, changed the course of history, heralding the rise of a new country that would grow into the current world power it is today. The unification of the 13 Colonies under one federal head made a huge difference in the victory that was achieved against England, and facilitated the offer of help from France. This also led to the creation of a House of Representatives, where citizens would elect their own leaders to speak for their communities in government. George Washington was an outstanding military commander and later the first President of the United States. This was, in part, because of his wisdom not to appear in the public as the new king of the country that had just waged war against England to remove the last one. Instead, he was willing to be the temporarily elected leader, and to include people in his Cabinet who had differing points of view from his own in order to make the best executive decisions for the new country. It was important for the new United States to compete economically with the other more established nations of the time. Though the Revolutionary War was costly in terms of lives and property, America did succeed in freeing itself from the tyrannical taxations demanded by England and established a new, more representative form of government—with a president, not a king, in command of the executive branch.

2.   The Barbary Pirates war that President Thomas Jefferson launched against the Barbary states (Tunis, Algeria, Morocco, and Tripoli [now Libya]) involved the United States battling Islamic terrorism at sea. These pirates, who came from the aforementioned countries, claimed their bounty in the name of Islam, while holding captured prisoners for ransom until their outrageous demands were met. In the case of President Thomas Jefferson, instead of continuing to pay these unfair ransoms, he had the bold initiative to defeat these pirates in their homeland, which restored American shipping to a safe and profitable enterprise that was free from terror. The United States literally won the right, through war, to freely navigate the seas, which was crucial to defeating tyranny at sea and gaining respectability on the world stage. President Jefferson is credited with this important military victory. Just as important, the U.S. Marine Corps was created during this time.

3.   Jefferson's successor, President James Madison, was heavily involved in warfare during his time in office in a rematch against former Revolutionary War opponent England. At the time, the United States was having problems with both England and France: American shipping was

being openly attacked on the high seas by the two nations. Yet it was England that had become America's number one aggressor, because England was inciting Native Americans, led by Tecumseh, to attack white settlers in America and was holding American naval personnel and commercial employees prisoner. In fact, many settlers were slaughtered by the British/Native American alliances, although not all Native American tribes were against the Americans. Madison asked Congress for a declaration of war against England on four counts.

1. Impressment of American seamen
2. Violation of American neutral rights
3. Blockading of U.S. ports
4. The British refusal to revise the orders in the Council of 1807, which barred all trade with France and her colonies

This costly war claimed a number of lives on both sides, and the nation's capital itself was mercilessly attacked by the British, who burned the Capitol building, the White House, the Library of Congress, and many other federal landmarks. The port of Baltimore was also attacked by the British. Despite these events, America prevailed a second time against its powerful English opponent and capitalized on commercial shipping free from impressment by England. Two additional results of this war were the Francis Scott Key composition of "The Star-Spangled Banner," after the British bombing of Fort McHenry in Baltimore, and the emergence of General and future President Andrew Jackson, who became a powerful hero after his complete victory against the British at the Battle of New Orleans. This would set the course of events that would lead to Jackson's rise to power, both as a general and a politician. In the case of President James Madison, he handled the presidential office admirably for two terms. Although circumstances thrust a difficult war on Madison when he was president, he led our nation to its very successful conclusion.

4. I have combined my analysis of every major war between the European, Caucasian-descended Americans and the Native Americans into one chapter. Although the Native Americans were involved in the American Revolutionary War and the War of 1812 on both sides, the wars mainly investigated in my chapter include the Creek War of 1813-14, the First and Second Seminole Wars, the Black Hawk War, and the Apache Wars. There were horrible massacres on both sides of these conflicts, including the Fort Mims Massacre that killed between 250 and 275 white settlers, the Trail of Tears forced migration of Native Americans from their homeland to the less fertile regions of Oklahoma, and the awful Sand Creek Massacre of 1864, which slaughtered 155 peaceful Native Americans from the Cheyenne and Arapaho tribes after they were rounded up by the white settlers. The massacre of Chief Big Foot's Sioux tribe at Wounded Knee in South Dakota was also analyzed. It is said that the Native American's biggest misfortune was that "they hold great bodies of rich lands," which seemed to summarize the European settlers' desire

to obtain those lands for cheap prices while maintaining levels of violence against displaced tribes. The first Seminole War brought both east and west Florida to the United States from Spanish and native control in 1818. It is now clear that despite any past differences, all Native American tribes must now be included and represented in our government, and more should be done to improve the quality of life and opportunities for all tribes of Native Americans. I have visited and studied at the National Museum of the American Indian here in Washington, D.C., and I have learned much about the plight and persecution of the Native American over the years. Studying this subject for my book opened my eyes to this controversial chapter in American history. I hope that all native tribes now present in the United States can live in peace and harmony with all American citizens.

5.   The Mexican War, which the United States fought from 1846 to 1848, did not seem to focus so much on the issue of terrorism, but instead that of land boundaries between the United States and Mexico. This dispute involved the concept of "Manifest Destiny," which then-President James Polk believed in, incorporating the theory that it was God's will for America to claim land all the way to the Pacific Ocean. This vision had developed into a problem with Mexico, because the boundaries in Texas were being disputed by both countries. After Polk's initial failure to buy the disputed land from the Mexican government, and the 2 massacres at the Alamo and Goliad, Polk felt justified to ask Congress for a declaration of war against Mexico, which he got. Polk was himself heavily influenced by former President Andrew Jackson. As a result of this war, the United States defeated Mexico and acquired the land that would become the states of Texas, New Mexico, and California.

6.   Perhaps the darkest chapter in American history occurred during the Civil War years of 1861 to 1865. This devastating conflict took over 620,000 lives on both sides. This war still ranks as the deadliest conflict in American history, and was fought over the controversial and morally wicked practice of the slavery of African-Americans in our society, with white slave owners considering and treating them as property. In 1860, newly elected President Abraham Lincoln and his northern-state constituents were opposed to the practice of slavery, while the southern states were heavily in favor of it. Although Lincoln was opposed morally and politically to the institution of slavery, in the beginning of his presidency, he was willing to allow it to continue where it already was, if the South would agree to remain in the Union. By Lincoln's inauguration day, the South did not agree. When the southern states began to systematically secede from the Union, Lincoln went to Congress for a formal declaration of war to fight the rebellious Confederate military forces, because he would not allow any states to secede from the country. Lincoln saw the evils of slavery and, thanks to his leadership, emancipated the slaves in the areas in rebellion with his proclamation in

1863. Lincoln also lived to see the official surrender of Confederate forces at Appomattox Courthouse in 1865. Lincoln employed African-Americans to serve in the Union Army during the war, where they performed admirably during the campaign. Lincoln's vision, bravery, and courage during this time literally saved the Union and made it a stronger entity. Although the South did score some important military victories in battle, arguably led by stronger military commanders like Generals Robert Lee and Stonewall Jackson, the South was still vastly outnumbered, and their Confederate government, led by Jefferson Davis, only had one political platform to work with, which further weakened that government in its fight against the Union. The Union, in contrast, operated under a multi-party system. At the end of this catastrophic chapter in American history, the Thirteenth, Fourteenth, and Fifteenth Amendments to the Constitution were added, outlawing slavery and giving African-Americans the right to vote. Although Lincoln himself was assassinated on April 14, 1865 at Ford's Theatre in Washington, D.C., he is remembered as one of our greatest presidents and one of the most successful leaders in American history. The traffic circles in the city of Washington, D.C. are named after Civil War heroes of the Union, including John Logan, Philip Sheridan, Winfield Scott, and Samuel Dupont.

7.   In the Spanish-American War that lasted from April 21st to August 12th, 1898, then-President William McKinley led our nation to war against an oppressive Spanish empire that had severely persecuted the Cuban people. In fact, one third of the Cuban people had died from Spanish rule there. McKinley, a Civil War veteran and former House member, had originally sought to avoid war with Spain, but circumstances, including the genocide of the Cuban people, failed diplomacy, and the destruction of the *U.S.S. Maine* led to his request to Congress for a declaration of war against Spain. The event of the *U.S.S. Maine*'s destruction in Havana harbor is similar to the September 11th, 2001 attacks in that it brought our nation together in a rallying cry of defeating the aggressors who had attacked America first and of freeing the Cuban people from such a cruel enemy. American troops fought and defeated Spain in both Cuba and the Philippines. Former Assistant Secretary of the Navy and Lieutenant Colonel Theodore Roosevelt gained valuable notoriety by displaying his bravery by resigning from McKinley's Cabinet to take his place on the battlefield, where he fought bravely and successfully. As a result, he was catapulted to the vice presidency as McKinley's running mate in 1900. In fact, he would succeed the very man who won the Spanish-American War as president because of an assassin's bullet in 1901. McKinley did succeed in getting an unconditional surrender from Spain, and the United States acquired Guam, the Philippines, and Puerto Rico as a result. Unfortunately and somewhat ironically, Cuba itself would fall to the Communists, led by Fidel Castro, in 1959.

8.   The conflict that would develop into the First World War began in July 1914 with the assassinations of Archduke Ferdinand and his wife, Sophie, in Serbia. This seemingly isolated incident exploded into unimaginable carnage involving 36 nations, on either the Allied side (Britain, France, Russia, and eventually the United States) or the Central Powers side (Germany, Austria-Hungary, the Ottoman Empire, and Bulgaria). This war was similar to the war versus terror after 9/11 because of the repeated German U-boat sinkings of commercial ships, including the *Lusitania*, which killed 1,198 people. There were new weapons of war used on the battlefield, including the machine gun, which was created during the Russo-Japanese War of 1905; poison gas; and Zeppelins that caused more destruction and death than had ever been seen on the battlefield. This also changed the dimensions of warfare itself. It brought a formidably dark cloud of destruction and mass death in its wake. Though President Woodrow Wilson wanted to keep the United States out of this global disaster and act as chief negotiator between these disputing countries, events would ultimately spin out of control, forcing Wilson early in his second term to rightfully ask Congress for a formal declaration of war against the Central Powers. Many lives were lost in this conflict, including many Russians, which led to the Bolshevik Revolution and the birth of Communism in the new Soviet Union. This completely ended Russia's involvement in the war. When this terrible conflict was over in November 1918, Wilson had tried, ill-fatedly, to create a League of Nations in order to avoid wars of this extreme magnitude in the future. However, both houses of Congress were controlled by the Republican Party, which voted against the idea of Wilson's League of Nations and wanted to return to a more isolationist policy in which the United States would not get involved in such global nightmares. This war also cost the lives of many American soldiers and many lives around the world. Wilson himself couldn't fix the dilemma due to the major stroke he had in office in 1919. Consequently, he was unable to complete the post-war process that could have prevented the next world war. Despite this, the U.S. armed forces made a difference in the Allied victory in World War One, and the United States emerged from the conflict as a world power.

9.   Due to America's isolationist policies after World War One, including its failure to sign the Treaty of Versailles, an over-punished Germany was overtaken by a new and more destructive regime called the Nazi Party, led by the evil Adolf Hitler. This new force unleashed a new wave of hostilities against its European neighbors such as had never been seen before. Hitler was originally from Austria, and did not even become a German citizen until the year he became chancellor, in 1933. That same year, newly elected President Franklin Roosevelt assumed power in America, and he promised to combat any new threats abroad and the severe economic depression at home. Roosevelt was President Wilson's Assistant Secretary of the Navy for his 2 terms in office. It took Roosevelt

some time to react not only to Hitler's aggression, but also to Japan's and Italy's. The U.S. military was downsized after World War One, so Roosevelt knew that it would take some time to rebuild. This was accomplished by President Roosevelt's New Deal programs that strengthened the American economy, and by sending people back to work, which reduced unemployment. Unfortunately, the Axis powers of Germany, Japan, and Italy were becoming more aggressive towards the rest of the countries in Europe and towards China, and the League of Nations was ill equipped to stop them. The Nazis opened their first concentration camp in Dachau during 1933, and their systematic annihilation of the Jewish people was well under way. This horrific act of evil that resulted in the deaths of over 6 million Jews is one that can never be forgotten or repeated. In the Pacific theatre, the Japanese military was behaving aggressively with its Asian neighbors, systematically murdering and raping its Chinese neighbors. When the attack on Pearl Harbor took place on December 7th, 1941, U.S. Navy personnel were surprised by the attacking Japanese aircraft. This attack killed 2,403 U.S. servicemembers and wounded an additional 1,110. This hostile act by a foreign government resulted in a similar death toll to that of the attack on 9/11, which also spurred our nation into war. When the United States became a joint partner of the Allied military coalition (United States, United Kingdom, and the Soviet Union), Roosevelt installed a series of successful economic programs that related to the production of war material. Roosevelt saw the wisdom in employing women in government jobs. Women also played a role in manufacturing military gear, equipment, and weapons, which increased the employment rate of women to 80%. The population in general saw a sharp decline in unemployment during the war years with Roosevelt in office. When President Roosevelt died in office on April 12, 1945, Harry Truman became the president and oversaw the successful conclusion to the war. At the time and until 1989, Germany was controlled by the Soviet Union east of Berlin, but the Soviets consequently relinquished their authority there in 1989 when their regime began to collapse. Japan became independent again in 1952. I witnessed a speech at the Library of Congress on December 7th, 2005, by the Japanese Ambassador to the United States. In the speech, the ambassador stated that the relationship between Japan and the United States has never been stronger. Japan and the United States have common values and enjoy great trade relations. Japan believes in freedom, democracy, and human rights. Japan's economy is currently ranked as number three in the world, and America is ranked number one. The two countries combine to make up approximately 40% of the world's gross national product, and Japan is America's staunchest ally in Asia. Normal and positive relations have also been restored with both Germany and Italy.

10. The Korean War, in which America was involved from June 25th, 1950 until July 26th, 1953, centered around the territorial dispute between

the Communist northern regime and the free, democratic south of Korea. The two countries are spilt at the 38th Parallel. At the end of World War Two, Korea was controlled by the Japanese and by the Soviet Union north of the 38th Parallel , and the south was controlled by the United States. The Communist north invaded the south on June 25th, 1950, and the war began on that day. President Harry Truman had vowed not to back down to the Communists, and his previous success in preventing the spread of Communism in Greece and Turkey in 1947 fueled his vision for Korea. In fact, Truman compared the North Korean aggression to that of both the Kaiser and Hitler. Truman was determined to fight it. After some initial success in Korea, moving the Communists back, Truman appointed General Douglas MacArthur to be the supreme commander in the field. MacArthur's threat to invade China after claiming all of Korea in battle provoked the Chinese military into the conflict, which cause massive casualties of U.S. soldiers. The capital city of Seoul was recaptured by the Communists in 1951, although the United States in turn recaptured Seoul shortly thereafter. Controversy erupted when Truman fired General MacArthur in 1951 for disobeying his orders, which is against America's Constitution. When President Dwight Eisenhower signed the armistice with North Korea on July 26th, 1953, the previous borders of the 38th Parallel, separating the Communist north from democracy in the south, were restored. However, this three-year war came at a very heavy price. There were 147,000 casualties for the U.S. military, with over 33,000 fatalities. The U.S. military still remains on the Korean peninsula, protecting the 38th Parallel to this very day.

    11.   The Vietnam War, which lasted for the United States from 1959 until 1973, was the only conflict in American history so far in which the military objectives were not met, and, accordingly, the war was not considered to be a success. The French military was also defeated previous to U.S. involvement, during 1946 to 1952. Before America's military involvement in Vietnam, President Dwight Eisenhower had funneled millions of dollars in aid to South Vietnamese leader Ngo Dinh Diem and his government in an effort to stop the Communist aggression there. While President John Kennedy's response of raising troop levels and financial commitments to the South Vietnamese government increased during his Administration, his overall plan was to have all American troops withdrawn from Vietnam by the end of 1965. This did not occur, because of the assassinations of not only Diem, on November 2, 1963, but also President Kennedy himself, on November 22, 1963. New President Lyndon Johnson, who didn't want to appear soft on Communism, escalated the number of troops in Vietnam during his Administration, from 15,000 at the end of 1963 to 540,000 by the end of 1968. Johnson's military policy and the lack of legitimate South Vietnamese leadership hurt his presidency, despite Johnson's push for his Great Society's agenda. Johnson's successor, Richard Nixon, sought to

gradually end American troop presence but shared the desire of his predecessors to keep Vietnam free from Communism south of the 17th Parallel. Nixon wanted to replace American troops with South Vietnamese troops. However, Nixon was widely criticized for invading Cambodia without Congressional approval, which caused widespread outrage and protests in America. The Kent State University incident, in which 4 students were killed on May 4, 1970 during a protest of the invasion of Cambodia, caused massive riots on 1,100 campuses across the nation. Nixon was trying to make diplomatic overtures to the Communist nations of China and the Soviet Union, but did not want to seem weak in dealing with the North Vietnamese. While Nixon was successful in negotiating important treaties with both China and the Soviets, he could not prevent the inevitable collapse of the South Vietnamese government and army, which eventually fell to the Communists in 1975. Nixon could not return to help South Vietnam's President Thieu after the U.S. withdrawal because he was consumed with the Watergate scandal, which led to his resignation from office in 1974. His successor, President Gerald Ford, did not resume hostilities with Vietnam, and, accordingly, the American political and military goals for the Vietnam War were not accomplished. The price for defeat in this war was heavy, with 58,151 U.S. soldiers killed and thousands more wounded. Although the country of Vietnam fell to Communist control in 1975, the fear of the remaining nations in Asia being overrun by Communist governments did not materialize, and thus the Communist regime in Vietnam was contained solely to that country. This war represented an ideological struggle, somewhat terror related, of freedom against Communism, which was prevalent during the Cold War years.

12.   Eighteen years after American forces withdrew from Vietnam, America was called to war again, this time in the Middle East, to liberate the country of Kuwait from the invading Republican Guard Army of Saddam Hussein's Iraq. This conflict began when Iraq invaded oil-rich Kuwait in August 1990. Consequently, President George H.W. Bush ordered Operation Desert Storm to begin on January 15th, 1991. This was President Bush's second military campaign during his term in office, the other one being the successful capture of Manuel Noriega and the liberation of the Panamanian people in December 1989. Bush felt that because of Hussein's unprovoked aggression against his Kuwaiti neighbors, other oil-rich countries in the region, most notably Saudi Arabia, would be next on Hussein's hit list for open conquest. This very real threat prompted Bush into action, and he successfully enlisted 23 countries in a military coalition to help expel Hussein and his army from Kuwait. The military Operations Desert Shield and Desert Storm were led by two capable generals, Colin Powell and Norman Schwarzkopf, who were then the Joint Chiefs of Staff Chairman and U.S. Central Command commander, respectively. These military leaders capitalized the strength

of America's military arsenal and capabilities, which can be attributed to the Cold War years of massive defense spending and build-up. It was clear to President Bush that Hussein, once an ally to the United States under the Reagan Administration, was now a direct threat to the Middle East region. This action by Hussein dictated a military response from a united coalition of nations that punished this unprovoked act of aggression. President Bush sent a total of 541,000 U.S. troops into the war theatre to liberate Kuwait. Operations Desert Storm and Desert Shield were a complete success, with minimal casualties and fatalities. However, knowing that he did not have a U.N. mandate to enter the country of Iraq to dispose of Hussein, President Bush declared the liberation of Kuwait a military success and, accordingly, ordered all U.S. troops home by January 1992. It would take another Bush presidency, that of his son, President George W. Bush, to invade Iraq and topple Hussein's regime 12 years later.

The 9/11 Commission Report says that America's fight with terror should be a global coalition strategy. It is important for the U.S. government to make substantial efforts in cooperation with foreign governments abroad to combat international terrorism. Monarchical governments still exist in Jordan, Saudi Arabia, and Morocco. What more can be done to better protect our citizens and leaders from terrorist attacks in the wake of September 11th, 2001? In fall 2001, President George W. Bush and Congress passed both the Aviation and Transportation Security Act and the Patriot Act. This was done to better prevent further acts of aggression by terrorists and those who wish to do us harm. The Patriot Act consists of 342 pages. President Bush also created the Homeland Security Department. All airports have been substantially upgraded since the 9/11 attacks on our nation to comply with all federal measures to prevent any possible attacks on airplanes or airports in general, which remains a constant threat in today's world.

General Anthony Zinni, the former head of U.S. Central Command, says in his book *The Battle for Peace* that it would be wise to integrate more federal agencies so that more intelligence can be readily available to prevent attacks in the future (#13). Accordingly, the Iraq Study Group, co-chaired by James Baker and Lee Hamilton, recommends a bi-partisan approach for new and enhanced diplomatic and political efforts that would better stabilize Iraq. The group also suggests that it is important for the Sunni and Shia populations in Iraq to reconcile. A key to stabilization in the Middle East is the relationship of Iraq with its immediate neighbors, especially Iran. Iran and the United States have had growing tensions in recent years, despite Iran supporting America's invasion of Afghanistan right after the September 11th, 2001 terrorist attacks. Despite this, the Iraq Study Group says that "of all its neighbors, Iran has the most leverage in Iraq" (#12).

The United States has a great interest in solving the Israeli-Palestinian conflict, and President George W. Bush made a commitment to establishing a two-state solution (Israel and Palestine) in June 2002. It is of equal importance for Afghanistan to remain free of terror groups and to support its democratically elected government. To help fight terrorism at home, the 9/11 Commission recommends a biometric system of screening at airports to allow speedier background checks, and more accurate intelligence on terror groups. The committee further recommends tightening security at sea ports across our nation's northern and southern borders with Canada and Mexico, respectively. The committee explains that the best intelligence, obtained both internationally and domestically, wins wars (#1). As General Zinni says, "we understand war making far better than peace making; we must make the effort to promote stability and peace. In that way, we will best promote the values we hope to share with those people in the world who suffer from its absence" (#13). Zinni also states that "we already know that the old ways of thinking about the world no longer apply," and that it will take real change and skilled diplomacy to adapt to this ever-changing globe of potential problems (#13).

According to Arthur Schlesinger's book, *War and the American Presidency*, he states that presidents can pick up a good idea or two from those who are dissenters. After all, Schlesinger states that "war presidents have never been exempt from criticism and dissent" (#15). He further states that "presidents are never infallible." Schlesinger points to a statement by former President Theodore Roosevelt in 1918: "To announce that there must be no criticism of the president, right or wrong, is not only unpatriotic and servile, but it is morally treasonable to the American public." An example of this healthy debate of conflicting political ideologies can be made from the vehement criticism Abraham Lincoln made public when he was a Congressman from Illinois about the Mexican War, which was an unpopular war at the time with American citizens.

In General and former Secretary of State Colin Powell's book, *My American Journey*, he expresses the desire not to unnecessarily risk soldiers' lives just for any campaign. Instead, Powell suggests that we must risk soldiers' lives only for worthy military objectives. He goes on to say, "if the duty of the soldier is to risk his or her life, the responsibility of their leaders is not to spend that life in vain" (#18). Powell goes on to say that "war should be the politics of last resort." In fact, it was former Defense Secretary Caspar Weinberger who recommended six tenets in committing troops abroad in any situation:

1. Commit only when our vital interests are at stake
2. If we commit, do so with all the resources necessary to win
3. Go in with clear military objectives
4. Be ready to change the commitment if the objectives change, because wars rarely stand still

5.   Only take on commitments that can gain the support of the American people and Congress

6.   Commit troops only as a last resort

It is clear, as Dr. Henry Kissinger states in his book *Does America Really Need a Foreign Policy?* that it is a number one priority for America to do everything possible to prevent the spread of nuclear technology worldwide. Speaking of the global stage, Kissinger suggests that it is important for America to work with Asian nations to find common goals for peace, and to continue our strong friendship with Japan. Kissinger says that Africa has the lowest existing economic rate in the world, with disease, civil war, and extreme corruption still as rampant as ever in that location (#3). Seventy percent of all A.I.D.S. cases worldwide originate in Africa. It is important to help and assist these poorer nations with international efforts to control these human hardships, and to do our part to implement humanitarian intervention and improve as many lives as possible around the world. Arthur Schlesinger says that the United States must work more congruently with the United Nations. "For all its defects, a world without the U.N. would be considerably more troublesome." Schlesinger points out that the biggest threat to the 21st century is global fanaticism (#15).

On November 8th, 2016, in a stunning upset, Republican candidate Donald Trump beat his Democratic opponent Hillary Clinton to become the 45th President of the United States. In his inauguration speech, he described the definition of American carnage by saying, "Mothers and children trapped in poverty in our inner cities, rusted-out factories like tombstones across the landscape of our nation. An education system flush with cash, but which leaves our young and beautiful students deprived of knowledge, and the crime, gangs, and drugs that have stolen too many lives and robbed our country of so much unrealized potential. From this day forward, a new vision will govern our land. From this moment on, it's going to be America first.... We will bring back our jobs, we will bring back our borders, we will bring back our wealth, and we will bring back our dreams. We will reinforce old alliances and form new ones, and unite the civilized world against radical Islamic terrorism, which we will eradicate completely from the Earth. When America is united, America is totally unstoppable. There should be no fear, we are protected and we will always be protected" (#17). As North Korean leader Kim Jong-un tests and threatens his neighbors with intercontinental ballistic missiles, with the unstable situation in the Middle East, notably in Syria and Iraq, and with increasing tensions between America and the Russian government, the threat against the United States and our allies is ever evolving. Trump did successfully launch 59 Tomahawk missiles at a Syrian air base that was responsible for making Sarin gas that was used against Syrian people, which killed many, and he did send a message to the international community that despite his inauguration speech of America first, he was still willing to respond militarily to any abuse of power internationally when circumstances dictate such action. However, Trump's ongoing domestic battles over health care and banning people from predominantly Muslim countries in the early months of his presidency have shown him the realities of how difficult the job of commander-in chief is, and only time will tell how he and his eventual successors handle the daunting responsibility of preventing our enemy from attacking our citizens again, as they horrifically did on September 11th, 2001.

In summary, each of the presidents I have analyzed in this book were faced with the same choices and consequences that President George W. Bush faced on September 11th, 2001, when he decided to take our nation to war against a foreign aggressor. Will we succeed in stopping terror from ever reaching our shores again, or will our enemy find new ways to attack our citizens and way of life? Will terrorism, or groups affiliated with terror, become our nation's new enemy for the foreseeable future? These are the questions that our commanders in the field, our political leaders, and our citizens should be contemplating every day, and the goal is to discover new ways of mutual cooperation to prevent 9/11-style attacks from ever happening again. Schlesinger says that "knowledge of yesterday provides guidance for tomorrow.... One cannot doubt that the study of history makes people wiser" (#15).

I have studied how former presidents dealt with their individual situations when they called our nation into war, and the results they achieved during their terms in office. I pray that one day, global conflicts can be settled in peaceful and non-violent ways. Until then, it is up to the President of the United States to use his or her best judgment to work with Congress, Cabinet officials, diplomats, and foreign governments when the United States goes to war, and to best protect our soldiers and citizens. As President Jimmy Carter said when he accepted the Nobel Peace Prize in 2002, "War may sometimes be a necessary evil, but no matter how necessary, it is always evil." May God bless our leaders, our troops, and our citizens. We will never forget those we lost on September 11th, 2001 or our troops, both past and present, who have made the ultimate sacrifice for our nation.

Bibliography

1. *The 9/11 Commission Report* (previously listed).
2. The U.S. Constitution.
3. *Does America Really Need a Foreign Policy? Toward a Diplomacy for the 21st Century*, by Dr. Henry Kissinger (Touchstone Books, 2001).
4. *Winning the Future: A 21st Century Contract With America*, by Newt Gingrich (Regency Publishing, 2005).
5. *Abraham: A Journey to the Heart of Three Faiths*, by Bruce Feiler (HarperCollins, 2002).
6. *God*, edited by Jacob Neuser (Pilgrim Press, 1997).

7. *The Story of Christianity: A Celebration of 2000 Years of Faith*, by Father Michael Collins and Matthew Price (D.K. Publishing, 1999).

8. *The House: The History of the U.S. House of Representatives*, by Dr. Robert V. Remini (Smithsonian Books, 2006).

9. *Betting on America: Why the U.S. Can Be Stronger After September 11th*, by James A. Cortada and Edward Wakin (Pearson Education, 2002).

10. *War of Nerves: Chemical Warfare From World War One to Al-Qaeda*, by Jonathan B. Tucker (Pantheon Books, 2006).

11. *State of Denial: Bush at War Part 3*, by Bob Woodward (Simon & Schuster, 2006).

12. *The Iraq Study Group Report: The Way Forward—A New Approach*, by co-chairs James A. Baker and Lee Hamilton (Vintage Books, 2006).

13. *The Battle for Peace: A Frontline Vision of America's Power and Purpose*, by General Anthony Zinni and Tony Koltz (Palgrave MacMillian, 2006).

14. *Character Is Destiny: Inspiring Stories Every Young Person Should Know and Every Adult Should Remember*, by Senator John McCain, with Mark Salter (Random House, 2005).

15. *War and the American Presidency*, by Arthur Schlesinger, Jr. (W.W. Norton, 2004).

16. "In the Rise of ISIS, No Single Missed Key, but Many Strands of Blame," by Ian Fisher (*New York Times*, November 18th, 2015).

17. President Donald Trump's inauguration speech, January 20, 2017.

18. *My American Journey*, by General Colin Powell and Joseph E. Persico (previously listed).

19. *The Koran*, based on the original English translation by J.M. Rodwell (Ballantine Books, 1993).

Additional Bibliography

1. *Dark Horse: The Surprise Election and Political Murder of President James A. Garfield*, by Kenneth D. Ackerman (Carroll and Graf, 2003), disc 7.

2. *The American Presidents Series: John Quincy Adams*, by Robert V. Remini (Times Books, 2002), disc 6.

3. *Oval Office: Stories of Presidents in Crisis From Washington to Bush*, edited by Nathaniel May (Adrenaline Books, 2002).

4. *The Audacity of Hope: Thoughts on Reclaiming the American Dream*, by Barack Obama (Crown Publishers, 2006).

5. *Presidential Courage: Brave Leaders and How They Changed America, 1789-1989*, by Michael R. Bechloss (Simon & Schuster, 2007).

About the Author

Thomas P. Athridge was born and raised in the Washington D.C. metropolitan area. He graduated with a bachelor's degree from Curry College in Milton, MA, where he majored in communications and minored in political science. He is currently an employee at the Library of Congress, since December 1998, working in the U.S. Anglo-American Acquisitions Division. Thomas has a keen interest in politics in both the past and present, and he is quite enthusiastic about the current political climate and about politics in general. Thomas currently resides in Bethesda, MD.

www.ingramcontent.com/pod-product-compliance
Lightning Source LLC
Chambersburg PA
CBHW021219090426
42740CB00006B/291